MATH FOR DISTRIBUTION SYSTEM OPERATORS

AF703581

MATH FOR DISTRIBUTION SYSTEM OPERATORS

Practice Problems to Prepare for Distribution System Operator Certification Exams

John Giorgi

FIRST EDITION

American Water Works Association

Disclaimer

Although this study guide has been extensively reviewed for accuracy, there may be an occasion to dispute an answer, either factually or in the interpretation of the question. Both AWWA and the author have made every effort to correct or eliminate any questions that may be confusing or ambiguous. If you do find a question that you feel is confusing or incorrect, please contact the AWWA Publishing Group.

Additionally, it is important to understand the purpose of this study guide. It does not guarantee certification. It is intended to provide the operator with an understanding of the types of math questions he or she will be presented with on a certification exam and the areas of knowledge that will be covered. AWWA highly recommends that you make use of the additional resources listed at the end of this study guide in preparing for your exam.

Copyright © 2007 American Water Works Association
All rights reserved
Printed in the United States of America
ISBN-10 1-58321-455-0
ISBN-13 978158321455-8

Library of Congress Cataloging-in-Publication Data

Giorgi, John.

Math for water distribution operators / John Giorgi and prepared by the editors of American Water Works Association. -- 1st ed.

p. cm.

Includes bibliographical references.

ISBN 1-58321-455-0

1. Water--Distribution--Mathematical models. 2. Water--Distribution--Problems, exercises, etc. 3. Engineering mathematics--Formulae. I. Title.

TD482.G559 2007
628.1'440151--dc22

2006051778

American Water Works Association

6666 West Quincy Avenue
Denver, Colorado 80235-3098
303.794.7711

This work is dedicated to my wife, Flora Zhou Giorgi; my children, Sara, Stephanie, and Steve; my mother, Thelma Giorgi; and my father, Albert Peter Giorgi.

CONTENTS

TABLES

ACKNOWLEDGMENTS

I would like to thank the staff and editors of the American Water Works Association, Mary Kay Kozyra, Colin Murcray, John Stubbart, and Bill Lauer; and the reviewers of the manuscript: Kevin Linder, Wemlinger Water Treatment Plant, Aurora, Colo.; Sid Copeland, City of Louisville Water Plant, Louisville, Colo.; and Ron Mathis, City of Golden, Colo.; for their help and guidance in making this book possible. Their assistance is greatly appreciated.

I would also like to thank fellow water treatment operators at El Dorado Irrigation District in California: Kurt Mikkola, William Petterson, Radenko Odzakovic, Robin Peck, Dave Herrmann, Aaron Davis, Zol Whitman, Jon Beale, Mike Royere, Jake Maker, Dean Bugher, Michael Ragan, Rick Talone, and Boyd Haycock.

I am grateful to my wife Flora and my children, Steve, Stephanie, and Sara. Their patience and support for my long hours working on this book is greatly appreciated.

PREFACE

The first edition of *Math for Distribution System Operators* was written to provide students and operators with examples of a variety of different problems that will be encountered both on certification exams and on the job. This book is divided into four parts consisting of material associated with the first four grade levels for distribution. The problems increase in difficulty as the grade number increases.

Each problem is presented with easily followed steps and comments to facilitate understanding. One possible way to go through the math problems presented in this study guide is for you to cover the page you are working on with a piece of paper or cardboard. Then slowly move the cover down until you can read the question. Do the problem on a separate piece of paper. Uncover the worked solution and compare your method and result to the book's method and result. If your answer is the same, but your method is different, that's okay. Remember that there may be more than one way to solve a problem. If there is a certain problem that gives you trouble, try to do the problem again on another day until you completely understand it. Do similar problems that may be found in the other grades in the book. The more math problems you do, the more comfortable you will become with them.

Included after each grade level are tests to help you determine where your strengths and weaknesses are. Each test consists of randomly chosen problems from the associated problems in that chapter. The test problems are followed immediately by the solution. You can complete the problem on a separate piece of paper. After completing the test, the procedures and answers can be checked against the provided procedures and solutions.

Common conversion factors are included in Appendix A for reference purposes and for doing problems in this book. Appendix B is a summary of the formulas used in this book. Appendix C consists of chemistry tables that are needed for some of the problems. Appendix D lists the US Environmental Protection Agency maximum contaminant level (MCL) tables for reference. And Appendix E lists the abbreviations used in this book.

Any suggestions for improving this math book, including additional types of problems, would be appreciated by the author. Please send your suggestions or questions to John Giorgi at aujourney@hotmail.com or in care of AWWA, 6666 Quincy Avenue, Denver, CO 80235-3098.

INTRODUCTION

SIGNIFICANT FIGURES

When you see an answer to a mathematical problem, laboratory test result, or other measured values, do you ever wonder how accurate they are? The accuracy of any answer is based on how accurate the values are in determining the answer or the accuracy of the laboratory result depends on the precision of the measuring instruments, and even the laboratory analyst.

The following discussion will show how to determine the number of significant figures or digits an answer to any particular problem should have, i.e., how many decimal places, if any, should the answer have.

The number 30.03 has four significant figures, while the number 33,000 has only two. Why is this so? The number 30.03 has been measured to the hundredth place so the zeroes that are straddled by the threes are significant. In fact, all figures to the left of a decimal point are significant (for example, 7.000 has 4 significant figures). The second number, 33,000 has only two because the zeroes are only placeholders and are thus not significant. See exercise below on significant figures.

"Rounding Off"

"Rounding off" numbers is simply the dropping of figures starting on the right until the appropriate numbers of significant figures remain. Let's look at the three rules and an example for each that governs the process of rounding numbers.

1. When a figure less than five is dropped, the next figure to the left remains unchanged. Thus, the number 11.24 becomes 11.2 when it is required that the four be dropped.
2. When the figure is greater than five that number is dropped and the number to the left is increased by one. Thus 11.26 will become 11.3.
3. When the figure that needs to be dropped is a five, round to the nearest even number. This prevents rounding bias. Thus 11.35 becomes 11.4 and 46.25 becomes 46.2.

The Significance or Insignificance of Zero

A zero may be a significant figure, if it is a measured value, or be insignificant and serve only as a spacer for locating the decimal point. If a zero or zeroes are used to give position value to the significant figures in the number, then the zero or zeroes are not significant. An example of this would be the following expression: 1.23 mm = 0.123 cm = 0.000123 m = 0.00000123 km. The zeroes are insignificant and only give the significant figures, 1 2 3, a position that dictates their value.

Addition and Subtraction

In addition and subtraction, only similar units and written to the same number of decimal places may be added or subtracted. Also, the number with the least decimal places, and not necessarily the fewest number of significant figures, places a limit on the number that the sum can justifiably carry. For example when adding the following numbers: 446 mm + 185.22 cm + 18.9 m.

First, convert the quantities to similar units, which in this case will be the meter (second row below). Next, choose the least accurate number, which is 18.9. It has only one number to the right of the decimal so the other two values will have to be rounded off (third row below).

446 mm	=	0.446 m	=	0.4 m
185.22 cm	=	1.8522 m	=	1.8 m
18.9 m	=	18.9 m	=	18.9 m
				21.1 m

When adding numbers (including negative numbers), the rule is that the least accurate number will determine the number reported as the sum. In other words, the number of significant figures reported in the sum cannot be greater than the least significant figure in the group being added. Another example is given below in which the least accurate number, 170, dictates how the other three numbers will have to be changed before addition is done:

1.023 g	=	1 g
23.22 g	=	23 g
170 g	=	170 g
1.008 g	=	1 g
195.251 g		195 g

However, you cannot report either of these values. The third value, 170 g, has two significant numbers, while all the others have four. The limiting factor is this third value, 170 g. The number 195 has three significant figures, and thus, cannot be used either. The answer must be reported as 200 g even though this looks wrong because it only has one significant figure!

Multiplication and Division

In multiplication or division, the number that has the fewest significant figures will dictate how the answer will be written. Suppose we had a problem where we had to multiply two numbers; $(23.88)(7.2) = 171.936$. The first number has 4 significant figures, while the second has only two. The answer should only be written with two significant figures, as 170, because one of the numbers, 7.2, has only two significant figures. In both multiplication and division, "rounding off" never should be done before the mathematical exercise. Only the result should be appropriately "rounded off"

In the following exercise, give the number of significant figures from each of the values below:

	VALUE	ANSWER
a.	8.34 lb/gal	3
b.	0.04 ntu	1
c.	23.06 mgd	4
d.	5 1-ton chlorine containers	infinite
e.	5.50 mg/L	3
f.	0.00000254	3
g.	50,000 tons	1
h.	6.023×10^{23} (the number of atoms in a mol, Avogadro's number)	4
i.	43,560 ft^3/acre-ft	infinite
j.	340 m	2
k.	5.01 g	3
l.	1,200,050 gal	6
m.	1,000,000/M	infinite
n.	1,440 min/d	infinite
o.	7.481 gal/ft^3	4
p.	86,400 s/d	infinite

ROUNDING IN THIS BOOK

The problems in this book are done in steps so the student can see each operation visually, which facilitates understanding. In so doing, a certain amount of "pre-rounding" has to occur, otherwise the numbers continue in most cases to absurdity. This "pre-rounding" was done in as much as possible such that the final answer would not be affected.

SET UP OF PROBLEMS IN THIS BOOK

Note that many of the same problem types have the equation written each time and that the units are written throughout each problem. The reason this is done is to help the student form a good habit in writing down the equations and units. This habit should be carried over during a certification test. In some cases partial credit will be given for a correct written equation. When the units are written down, it will not only help the student in setting up the problem correctly by seeing how the units will cancel, but will also make it easier for the person correcting the test to see your intent and also in correcting the problems.

Pi (π)

The number used for pi (π) will be 3.14 throughout this book.

DERIVATION OF THE NUMBER 0.785

The number 0.785 is used extensively in this book in conjunction with the diameter squared. Examples include determination of the area of a circle or the volume of a tank. The same answer can be achieved using πr^2, where "r" equals the radius. But how was this number, 0.785, derived?

The following applies:

1. **(0.785)(Diameter)^2 = πr^2**
2. **Diameter^2 = 4r^2**

Assume x is this number (0.785) and is not yet known. We know some number, x, times the diameter squared equals π times the radius squared. The equation is:

$$x(\text{Diameter})^2 = \pi r^2$$

From 1 above and substituting x for (0.785),

substitute 4r^2 for Diameter^2 from 2 above:

$$x(4r^2) = \pi r^2$$

Solve for x,

$$x = \frac{\pi r^2}{4r^2} = \frac{\pi}{4}$$

Substitute 3.14 for π and solve:

$$x = \frac{3.14}{4} = \mathbf{0.785}$$

Thus, x is equal to 0.785, which is what we wanted to prove.

Most water treatment operators like to use (0.785)(Diameter)^2, while engineers and scientists like to use πr^2. Because both will be encountered, it is advisable to know both methods.

CHAPTER 1

DISTRIBUTION

Grade 1

Students preparing for grade 2 through grade 4 distribution certification tests should also understand these problems.

CONVERSION PROBLEMS

1. Convert 723 gal to liters.

Number of L = (723 gal)(3.785 L/gal) = 2,737 L, round to **2,740 L**

2. Convert 81.0 ft^3 to gallons and liters.

First, find the number of gallons.

Number of gal = (Number of ft^3)(7.48 gal/ft^3)

Number of gal = (81.0 ft^3)(7.48 gal/ft^3) = 605.88 gal, round to 606 gal

Now, determine the number of liters.

Number of L = (81.0 ft^3)(7.48 gal/ft^3)(3.785 L/gal) = 2,293.26 L, round to **2,290 L**

3. Convert 56 ft^3/s to gallons per minute.

Number of gpm = (56 ft^3/s)(60 s/min)(7.48 gal/ft^3)
= 25,132.8 gpm, round to **25,000 gpm**

4. Convert 23 in. to centimeters.

Number of cm = (23 in.)(2.54 cm/in.) = 58.42 cm, round to **58 cm**

5. Convert 1.98 acres to square feet.

Number of ft^2 = (1.98 acres)(43,560 ft^2/acres) = 86,248.8 ft^2, round to **86,200 ft^2**

6. Convert 2,367 g to pounds.

Number of lb = (2,367 g)(1 lb/454 g) = 5.2136 lb, round to **5.21 lb**

7. Convert 3.45 mgd to cubic feet per second.

Number of ft^3 = (3.45 mgd)(1,000,000/1 M)(1 ft^3/7.48 gal)(1 d/86,400 s)
= **5.34 ft^3/s**

8. Convert 8.77 acre-ft to gallons.

$$\text{Number of gal} = (8.77 \text{ acre-ft})(43{,}560 \text{ ft}^3/\text{acre-ft})(7.48 \text{ gal/ft}^3)$$
$$= 2{,}857{,}518.6 \text{ gal, round to } \mathbf{2{,}860{,}000 \text{ gal}}$$

9. Convert 56 grams per gallon to milligrams per liter.

$$\text{Number of mg/L} = \frac{(56 \text{ gpg})(17.12 \text{ mg/L})}{1 \text{ gpg}} = 958.72 \text{ mg/L, round to } \mathbf{960 \text{ mg/L}}$$

10. Convert 1,472 L to gallons.

$$\text{Number of gal} = (1{,}472 \text{ L})(1 \text{ gal}/3.785 \text{ L}) = \mathbf{388.9 \text{ gal}}$$

11. Convert 8,711,400 gal to cubic feet and acre-feet.

First, find the number of cubic feet.

$$\mathbf{Number\ of\ ft^3} = \frac{\mathbf{(Number\ of\ gal)}}{\mathbf{(7.48\ gal/ft^3)}}$$

$$\text{Number of ft}^3 = \frac{(8{,}711{,}400 \text{ gal})}{(7.48 \text{ gal/ft}^3)} = 1{,}164{,}625.7 \text{ ft}^3\text{, round to } \mathbf{1{,}160{,}000 \text{ ft}^3}$$

Note: 7.48 gal/ft^3 defines the number of significant figures and thus is the reason we have to drop 4,625.7 ft^3 off the answer.

Next, find the number of acre-feet using the unrounded number.

$$\text{Number of acre-ft} = \frac{(\text{Number of ft}^3)}{(43{,}560 \text{ ft}^3/\text{acre-ft})}$$

$$\text{Number of acre-ft} = \frac{(1{,}164{,}625.7 \text{ ft}^3)}{(43{,}560 \text{ ft}^3/\text{acre-ft})} = 26.74 \text{ acre-ft, round to } \mathbf{26.7 \text{ acre-ft}}$$

(continued)

Alternatively, we could have used the following formula:

$$\text{Number of acre-ft} = \frac{(\text{Number of gal})}{(43{,}560\ \text{ft}^3/\text{acre-ft})(7.48\ \text{gal/ft}^3)}$$

$$\text{Number of acre-ft} = \frac{(8{,}711{,}400\ \text{gal})}{(43{,}560\ \text{ft}^3/\text{acre-ft})(7.48\ \text{gal/ft}^3)} = 26.736\ \text{acre-ft, round to } \mathbf{26.7\ acre\text{-}ft}$$

PERCENT CONVERSIONS

12. Convert 56.3% to decimal form.

$$\frac{56.3\%}{100\%} = \mathbf{0.563}$$

13. Convert 45.23% to decimal form.

$$\frac{45.23\%}{100\%} = \mathbf{0.4523}$$

14. Convert 1.058 to percent.

$$(1.058)(100\%) = \mathbf{105.8\%}$$

TEMPERATURE CONVERSIONS

15. Convert 17°C to degrees Fahrenheit.

The equation for Fahrenheit is:

$$°F = (9/5)°C + 32$$

$$°F = (9/5)17°C + 32 = 62.6°F, \text{ round to } \mathbf{63°F}$$

16. Convert 451°F to degrees Celsius.

The equation for Celsius is:

$$°C = 5/9(°F - 32)$$

$$°C = 5/9(451°F - 32) = 232.77°C, \text{ round to } \mathbf{233°C}$$

17. Convert –34°C to degrees Fahrenheit.

$$°F = (9/5)-34°C + 32 = -29.2°F, \text{ round to } \mathbf{-29°F}$$

18. Convert 212°F to degrees Celsius.

$$°C = 5/9(212°F - 32) = \mathbf{100°C}$$

PERCENT AND PARTS PER MILLION CONVERSION PROBLEMS

19. Convert a chlorine solution of 2.5 ppm to percent.

$$\text{A 1\% solution} = 10{,}000 \text{ ppm or } 10{,}000 \text{ mg/L}$$

$$\frac{2.5 \text{ ppm}}{10{,}000 \text{ ppm}/1\%} = \mathbf{0.00025\% \text{ solution}}$$

An alternative way to convert 2.5 ppm to percent is:

$$\frac{2.5 \text{ ppm} \times 100\%}{1{,}000{,}000/1 \text{ M}} = \mathbf{0.00025\% \text{ solution}}$$

20. A solution was found to be 7.6% hypochlorite. How many milligrams per liter of hypochlorite are in the solution?

If a 1% solution has 10,000 mg/L, a 7.6% solution will have:

$$\frac{(7.6\%)(10{,}000 \text{ mg/L})}{1\%} = \mathbf{76{,}000 \text{ mg/L hypochlorite}}$$

or:

$$\frac{1\%}{7.6\ \%} = \frac{10{,}000}{x}$$

Then, solve for x.

$$1\%x = 7.6\%(10{,}000)$$

$$x = \frac{7.6\%(10{,}000)}{1\%} = \mathbf{76{,}000 \text{ mg/L hypochlorite}}$$

21. Convert 72 ppm to percent.

$$\frac{72 \text{ ppm} \times 100\%}{1{,}000{,}000/1 \text{ M}} = \mathbf{0.0072\%}$$

22. Convert 63.5% to parts per million.

$$\frac{(63.5\%)(1{,}000{,}000)}{(100\%)(1\text{ M})} = \mathbf{635{,}000\ ppm}$$

AVERAGE CALCULATIONS

23. An operator records her odometer reading after filling up her truck with gas as 002834.5. About a week later she buys 14.8 gal of gas and records the odometer reading as 003093.9. What was the truck's miles per gallon?

First, determine the number of miles between fill-ups.

$$\text{Number of mi} = 003093.9\text{ mi} - 002834.5\text{ mi} = 259.4\text{ mi}$$

Then, determine the miles per gallon by dividing the number of miles traveled by the number of gallons added.

$$\text{Number of mpg} = \frac{259.4\text{ mi}}{14.8\text{ gal}} = \mathbf{17.5\ mpg}$$

24. Records on a pump indicate that it pumped 1,018,200 gal for the month of May. What is the average number of gallons pumped per day?

$$\text{Number of gpd} = \frac{\text{Number of gallons pumped}}{\text{Number of days}}$$

$$\text{Number of gpd} = \frac{1{,}018{,}200\text{ gal}}{31\text{ d}} = \mathbf{32{,}845\ gpd}$$

25. The reading on an operator's truck odometer read 32,774.5 after he had filled his truck with 19.3 gal of gas. If the odometer read 32,497.9 the last time he got gas, what is the truck's miles per gallon?

First, determine the number of miles between fill-ups.

$$\text{Number of mi} = 32{,}774.5\text{ mi} - 32{,}497.9\text{ mi} = 276.6\text{ mi}$$

Then, determine the miles per gallon by dividing the number of miles traveled by the number of gallons added.

$$\text{Number of mpg} = \frac{276.6\text{ mi}}{19.3\text{ gal}} = \mathbf{14.3\ mpg}$$

26. Records for a distribution pump indicate the following:

Monday at 08:05 read is 0781844 gal × 100

Next day, Tuesday at 08:20 read is 0790107 gal × 100

How many gallons were pumped in gallons per minute assuming an average rate over the specified time period?

First, subtract the reads and multiply by 100 to find the number of gallons pumped.

$$\text{Number of gallons pumped} = (0790107\text{ gal} - 0781844\text{ gal})(100) = 826{,}300\text{ gal}$$

Then, find the number of minutes between the reads.

$$1\text{ d} + 15\text{ min} = 24\text{ hr}(60\text{ min/hr}) + 15\text{ min} = 1{,}455\text{ min}$$

Next, divide by the time in minutes.

$$\text{Number of gpm (average)} = \frac{826{,}300\text{ gal}}{1{,}455\text{ min}} = 567.9\text{ gpm, round to } \mathbf{568\ gpm\ (average)}$$

PERCENT CALCULATIONS

27. What percent is 12,887 of 475,258?

The equation is:

$$\textbf{Percent} = \frac{\textbf{(Low number)(100\%)}}{\textbf{High number}}$$

$$\frac{(12{,}887)(100\%)}{475{,}258} = \textbf{2.7116\%}$$

There are five significant figures in this problem.

28. Calculate 42.0% of 7,310.

The equation is:

$$\textbf{Percent} = \frac{\textbf{(Low number)(100\%)}}{\textbf{High number}}$$

The question is asking for the low number. Rearrange the equation to solve for the low number. The following will explain in detail how this is done. These operations are done throughout this book without the detail, so before going on with the rest of the problems, make sure you thoroughly understand these manipulations of the equations. Whatever you do to one side of the equation, you have to do to the other. In the example that follows, our goal is to get the low number by itself.

The first step is to multiply both sides by the High number.

$$\text{(Percent)(High number)} = \frac{\text{(Low number)(100\%)(High number)}}{\text{High number}}$$

On the right side of the equation, the high number in the numerator (top) cancels out the high number in the denominator (bottom).

Thus, we get:

$$\text{(Percent)(High number)} = \text{(Low number)(100\%)}$$

(continued)

Next, divide both sides of the equation by 100% to get the low number by itself.

$$\frac{\text{(Percent)(High number)}}{100\%} = \frac{\text{(Low number)(100\%)}}{100\%}$$

On the right side of the equation, 100% in the numerator (top) cancels out 100% in the denominator (bottom).

Thus, we get:

$$\frac{\text{(Percent)(High number)}}{100\%} = \text{Low number}$$

By convention we will put the low number on the left side of the equation to get:

$$\text{Low number} = \frac{\text{(Percent)(High number)}}{100\%}$$

Substitute known values and solve.

$$\text{Low number} = \frac{(42.0\%)(7{,}310)}{100\%} = \mathbf{3{,}070}$$

29. Twenty-two percent of what number is exactly 427.9? Calculate to the nearest unit.

We need to find the high number. As in the previous problem, rearrange the original equation given in problem 28 to solve for the high number.

$$\mathbf{Percent} = \frac{\mathbf{(Low\ number)(100\%)}}{\mathbf{High\ number}}$$

$$\text{High number} = \frac{\text{(Low number)(100\%)}}{\text{Percent}}$$

$$\text{High number} = \frac{(427.9)(100\%)}{22\%} = \mathbf{1{,}945}$$

30. Nineteen is what percent of 164?

The equation is:

$$\textbf{Percent} = \frac{\textbf{(Low number)(100\%)}}{\textbf{High number}}$$

$$\text{Percent} = \frac{(19)(100\%)}{164} = 11.6\%, \text{ round to } \mathbf{12\%}$$

SOLVING RATIOS AND DETERMINING PROPORTIONS

31. If alum usage at a water plant averages 812 gpd, how many gallons will the plant use in 30 days?

Set up a ratio and let x equal the unknown.

$$\frac{x \text{ gal}}{30 \text{ d}} = \frac{812 \text{ gal}}{1 \text{ d}}$$

Solving for x, multiply both sides by 30 d.

Note: The 30 d cancels out on the left side of the equation. For convenience, it is not shown here or throughout the rest of this book.

$$x \text{ gal} = \frac{(812 \text{ gal})(30 \text{ d})}{1 \text{ d}} = \mathbf{24{,}360 \text{ gal}}$$

32. In the following proportion, solve for the unknown value, x.

$$35{:}x = 12{:}60$$

Write the proportion in fraction form and solve for x.

$$\frac{35}{x} = \frac{12}{60}$$

$$\frac{(35)(60)}{12} = x$$

(continued)

or:

$$x = \frac{(35)(60)}{12} = \mathbf{175}$$

Thus, the proportion is:

$$\mathbf{35{:}175 = 12{:}60}$$

33. Solve the following proportion.

$$4{:}96 = x{:}768$$

Write the proportion in fraction form and solve for x.

$$\frac{4}{96} = \frac{(x)}{768}$$

$$x = \frac{(4)(768)}{96} = \mathbf{32}$$

Thus, the proportion is:

$$\mathbf{4{:}96 = 32{:}768}$$

34. If 50 lb of soda ash (one bag) costs $5.25, what is the cost of 60 bags assuming no discount for buying a larger quantity?

Set up a ratio and solve for the unknown, x.

$$\frac{(x)}{60 \text{ bags}} = \frac{\$5.25}{1 \text{ bag}}$$

$$x = \frac{(\$5.25)(60 \text{ bags})}{1 \text{ bag}} = \mathbf{\$315.00}$$

35. If chlorine is being feed at a rate of 125 lb/d for a flow rate of 25 ft³/s, what should the adjustment on the chlorinator be when the flow rate is increased to 32 ft³/s?

Set up a ratio and solve for the unknown, x.

$$\frac{x\text{, lb/d}}{32\text{ ft}^3\text{/s}} = \frac{125\text{ lb/d}}{25\text{ ft}^3\text{/s}}$$

$$x = \frac{(125\text{ lb/d})(32\text{ ft}^3\text{/s})}{25\text{ ft}^3\text{/s}} = \mathbf{160\ lb/d}$$

PERIMETER DETERMINATIONS

36. Find the perimeter of a square that measures 23 ft on one side.

Perimeter of square = 4(Length of one side)

Perimeter of square, ft = 4(23 ft) = **92 ft**

37. Determine the perimeter of a triangle in feet with the following dimensions.

Side 1 = 7.0 in. Side 2 = 4.0 in. Side 3 = 8.0 in.

First, add the lengths of all three sides.

Perimeter, in. = (7.0 in. + 4.0 in. + 8.0 in.) = 19 in.

Next, convert inches to feet.

Number of ft = (19 in.)(1 ft/12 in.) = 1.583 ft, round to **1.6 ft**

38. Find the perimeter of a tank that has a radius of 63.5 ft.

In this case the perimeter is the same as the circumference.

The equation is:

Perimeter (Circumference) = (2)(π)(Radius)

Note: In this book, pi (π) equals 3.14.

Circumference = (2)(3.14)(63.5 ft) = 398.78 ft, round to **399 ft**

39. What is the perimeter in feet of a filter that is 29.75 ft long and 18.2 ft wide?

Because this filter is rectangular in shape, add the length of the two sides given and multiply by two.

The equation is:

Perimeter, ft = 2(Length + Width)

Perimeter, ft = 2(29.75 ft + 18.2 ft) = 2(47.95 ft) = **95.9 ft**

CIRCUMFERENCE, RADII, AND DIAMETER DETERMINATIONS

40. What is the circumference of a tank that is 110.0 ft in diameter?

The equation is:

Circumference = (π)(Diameter)

Circumference = (3.14)(110.0 ft) = 345.4 ft, round to **345 ft**

41. The circumference of a tank is 392.5 ft. What is the diameter?

The equation is:

$$\textbf{Circumference} = \boldsymbol{(\pi)}\textbf{(Diameter)}$$

Rearrange the equation to solve for the diameter.

$$\text{Diameter} = \frac{\text{Circumference}}{(\pi)}$$

$$\text{Diameter} = \frac{392.5 \text{ ft}}{3.14} = \textbf{125 ft}$$

42. The radius of a circular concrete area is 42.5 ft. What is the circumference?

The equation is:

$$\textbf{Circumference} = \textbf{(2)}\boldsymbol{(\pi)}\textbf{(Radius)}$$

$$\text{Circumference} = (2)(3.14)(42.5 \text{ ft}) = 266.9 \text{ ft, round to } \textbf{267 ft}$$

43. The circumference of a tank is 78.5 ft. What is the diameter?

The equation is:

$$\textbf{Circumference} = \boldsymbol{(\pi)}\textbf{(Diameter)}$$

Rearrange the equation to solve for the diameter.

$$\text{Diameter} = \frac{\text{Circumference}}{(\pi)}$$

$$\text{Diameter} = \frac{78.5 \text{ ft}}{3.14} = \textbf{25.0 ft}$$

44. The diameter of a tank is 210 ft. What is the circumference?

The equation is:

$$\textbf{Circumference} = (\pi)(\textbf{Diameter})$$

$$\text{Circumference} = (3.14)(210\text{ ft}) = 659.4\text{ ft, round to } \textbf{660 ft}$$

PRESSURE PROBLEMS

45. What is the psi at the bottom of a tank if the water level is 28.14 ft deep?

The equation is:

$$\textbf{psi} = \frac{\textbf{Depth, ft}}{\textbf{2.31 ft/psi}}$$

$$\text{psi} = \frac{28.14\text{ ft}}{2.31\text{ ft/psi}} = 12.18\text{ psi, round to } \textbf{12.2 psi}$$

46. What is the depth of water in a lake if the psi is 56.7?

The equation is:

$$\textbf{psi} = \frac{\textbf{Depth, ft}}{\textbf{2.31 ft/psi}}$$

Rearrange and solve.

$$\text{Depth, ft} = (56.7\text{ psi})(2.31\text{ ft/psi}) = 130.977\text{ ft, round to } \textbf{131 ft deep}$$

47. What is the psi 12 ft above the bottom of a lake that is 142 ft deep?

First, subtract the number of feet above the bottom of the lake from the total depth.

Depth at point in question = 142 ft – 12 ft = 130 ft

The equation is:

$$\textbf{psi} = \frac{\textbf{Depth, ft}}{\textbf{2.31 ft/psi}}$$

Substituting:

$$\text{psi} = \frac{130 \text{ ft}}{2.31 \text{ ft/psi}} = 56.3 \text{ psi, round to } \textbf{56 psi}$$

48. If the water depth in a reservoir is 31.45 ft, what is the pressure at 25 ft below the surface in pounds per square foot?

The equation using pressure in lb/ft^2 is:

$$\textbf{Pressure, lb/ft}^2 = \textbf{(Depth)(Density, 62.4 lb/ft}^3\textbf{)}$$

$$\text{Pressure} = (25 \text{ ft})(62.4 \text{ lb/ft}^3) = 1{,}560 \text{ lb/ft}^2\text{, round to } \textbf{1,600 lb/ft}^2$$

49. If a lake is 315 ft deep, what is the psi on the bottom?

Another number to commit to memory is 0.433 psi/ft, which is 1 psi divided by 2.31 ft.

The equation is:

$$\textbf{psi} = \textbf{(Depth, ft)(0.433 psi/ft)}$$

$$\text{psi} = (315 \text{ ft})(0.433 \text{ psi/ft}) = 136.4 \text{ psi, round to } \textbf{136 psi}$$

50. What is the pressure in pounds per square inch on a SCUBA diver if she is 65.5 ft below a lake's surface?

The equation is:

$$\textbf{psi} = \textbf{(Depth, ft)(0.433 psi/ft)}$$

$$\text{psi} = (65.5\ \text{ft})(0.433\ \text{psi/ft}) = \textbf{28.4 psi}$$

51. What would the number of atmospheres be on the SCUBA diver due to the lake's water only in problem 50?

The following applies:

$$1\ \text{atm} = 14.7\ \text{psi}$$

$$\text{Number of atm} = \frac{28.4\ \text{psi}}{14.7\ \text{atm}} = \textbf{1.93 atm}$$

AREA PROBLEMS

52. Find the area in square feet for a rectangular-shaped sedimentation basin that is 392 ft in length and 71.5 ft in width.

The equation is:

$$\textbf{Number of ft}^2 = \textbf{(Length)(Width)}$$

$$\text{Number of ft}^2 = (392\ \text{ft})(71.5\ \text{ft}) = 28{,}028\ \text{ft}^2\text{, round to } \textbf{28,000 ft}^2$$

53. What is the area of a tank if the tank's diameter is 30.4 ft?

The equation is:

$$\textbf{Area} = \textbf{(0.785)(Diameter)}^2$$

Some operators like to use this formula:

$$\textbf{Area} = \pi r^2$$

Most operators prefer the former so it will be used here.

$$\text{Area of tank} = (0.785)(30.4 \text{ ft})^2$$

$$\text{Area of tank} = (0.785)(30.4 \text{ ft})(30.4 \text{ ft}) = 725.47 \text{ ft}^2\text{, round to } \mathbf{725\ ft^2}$$

54. A triangle has a height of 17 in. and a base of 22 in. What is its area in square inches?

The equation is:

$$\textbf{Area of triangle} = \frac{\textbf{(Base)(Height)}}{\textbf{2}}$$

$$\text{Area of triangle, in.}^2 = \frac{(17 \text{ in.})(22 \text{ in.})}{2} = 187 \text{ in.}^2\text{, round to } \mathbf{190\ in.^2}$$

55. What is the cross-sectional area of a canal that is shaped like a trapezoid if the bottom of the canal (b_2) is 8.5 ft, the top of the water across the canal (b_1) is 12.8 ft, and the depth of the water is 5.6 ft?

The equation is:

$$\textbf{Area of trapezoid} = \frac{\mathbf{(b_1 + b_2)}\textbf{(Height or depth)}}{\textbf{2}}$$

$$\text{Area of trapezoid} =$$

$$\frac{(12.8 \text{ ft} + 8.5 \text{ ft})(5.6 \text{ ft})}{2} = \frac{(21.3 \text{ ft})(5.6 \text{ ft})}{2} = 10.65 \text{ ft}(5.6 \text{ ft}) = 59.6 \text{ ft}^2\text{, round to } \mathbf{60\ ft^2}$$

56. What is the area of a tank if the tank's radius is 45.6 ft?

The equation is:

$$\textbf{Area} = (\pi)\mathbf{r}^2$$

$$\text{Area} = (3.14)(45.6\ \text{ft})(45.6\ \text{ft}) = 6{,}529\ \text{ft}^2\text{, round to } \mathbf{6{,}530\ ft^2}$$

57. Determine the cross-sectional area of a canal that is shaped like a trapezoid if the bottom of the canal (b_2) is 6.4 ft, the top of the water across the canal (b_1) is 10.7 ft, and the depth of water is 4.4 ft.

The equation is:

$$\textbf{Area of trapezoid} = \frac{(\mathbf{b_1} + \mathbf{b_2})(\textbf{Height or Depth})}{\mathbf{2}}$$

$$\text{Area of trapezoid} = \frac{(10.7\ \text{ft} + 6.4\ \text{ft})(4.4\ \text{ft})}{2}$$

$$= \frac{(17.1\ \text{ft})(4.4\ \text{ft})}{2} = 8.55\ \text{ft}(4.4\ \text{ft}) = 37.6\ \text{ft}^2\text{, round to } \mathbf{38\ ft^2}$$

58. If the tank's diameter is 92.5 ft, what is the area of the tank?

The equation is:

$$\textbf{Area} = (\mathbf{0.785})(\textbf{Diameter})^2$$

$$\text{Area of tank} = (0.785)(92.5\ \text{ft})(92.5\ \text{ft}) = 6{,}716.7\ \text{ft}^2\text{, round to } \mathbf{6{,}720\ ft^2}$$

VOLUME DETERMINATIONS

59. What is the volume of a trench in cubic feet if it is 245 ft in length, 4.2 ft in width, and 5.8 ft in depth?

The equation is:

$$\textbf{Volume, ft}^3 = \textbf{(Length)(Width)(Depth) or (L)(W)(D)}$$

Volume, ft^3 = (245 ft)(4.2 ft)(5.8 ft) = 5,968.2 ft^3, round to **6,000 ft^3**

60. What is the capacity of a tank in cubic feet if it has a diameter of 75.2 ft and the height is 42.3 ft from the base?

The equation is:

$$\textbf{Volume, ft}^3 = \textbf{(0.785)(Diameter)}^2\textbf{(Depth)}$$

Volume, ft^3 = (0.785)(75.2 ft)(75.2 ft)(42.3 ft) = 187,778 ft^3, round to **188,000 ft^3**

61. How many liters of zinc orthophosphate can be contained in a tank that has a diameter of 10.5 ft and can be filled to a height of 9.0 ft?

The equation is:

$$\textbf{Volume, L} = \textbf{(0.785)(Diameter)}^2\textbf{(Depth)(7.48 gal/ft}^3\textbf{)(3.785 L/gal)}$$

Volume, L = (0.785)(10.5 ft)(10.5 ft)(9.0 ft)(7.48 gal/ft^3)(3.785 L/gal)

= 22,053 L, round to **22,000 L**

62. If a trench is 346 ft long, 4.4 ft wide, and 5.7 ft deep, how many cubic yards of soil were excavated?

First, find the number of cubic feet.

The equation is:

$$\textbf{Volume, ft}^3 = \textbf{(L)(W)(D)}$$

$$\text{Volume, ft}^3 = (346 \text{ ft})(4.4 \text{ ft})(5.7 \text{ ft}) = 8{,}677.68 \text{ ft}^3$$

Next, convert cubic feet to cubic yards.

The following applies: 1 yd^3 = 27 ft^3.

Thus,

$$\frac{8{,}677.68 \text{ ft}^3}{27 \text{ ft}^3/\text{yd}^3} = 321.4 \text{ yd}^3\text{, round to } \textbf{320 yd}^3$$

63. A trench that is 156 ft long, 3.8 ft wide, and 5.8 ft deep fills with water. How many gallons are contained in the trench?

The equation is:

$$\textbf{Volume, gal} = \textbf{(Length)(Width)(Depth)(7.48 gal/ft}^3\textbf{)}$$

$$\text{Volume, gal} = (156 \text{ ft})(3.8 \text{ ft})(5.8 \text{ ft})(7.48 \text{ gal/ft}^3) = 25{,}718 \text{ gal, round to } \textbf{26,000 gal}$$

64. A reservoir is found to average 56 ft in depth. The shape of the lake is approximately circular with a diameter of approximately 5,710 ft. How many acre-feet of water does the lake contain?

The equation is:

$$\textbf{Volume, acre-ft} = \frac{\textbf{(0.785)(Diameter)}^2\textbf{(Depth)}}{\textbf{43,560 ft}^3\textbf{/acre-ft}}$$

$$\text{Volume, acre-ft} = \frac{(0.785)(5{,}710 \text{ ft})(5{,}710 \text{ ft})(56 \text{ ft})}{43{,}560 \text{ ft}^3\text{/acre-ft}}$$

= 32,903 acre-ft, round to **33,000 acre-ft**

65. A pipe is 16.0 in. in diameter and 550 ft long. How many gallons does the pipe contain?

First, convert the diameter from inches to feet.

$$\text{Number of ft} = \frac{16.0 \text{ in.}}{12 \text{ in./ft}} = 1.33 \text{ ft}$$

Next, calculate the volume using the following formula:

$$\textbf{Number of gal} = \textbf{(0.785)(Diameter)}^2\textbf{(Length)(7.48 gal/ft}^3\textbf{)}$$

$$\text{Number of gal} = (0.785)(1.33 \text{ ft})(1.33 \text{ ft})(550 \text{ ft})(7.48 \text{ gal/ft}^3)$$

= 5,713 gal, round to **5,700 gal**

FLOW PROBLEMS

66. What is the velocity of flow in feet per second for a 10.0-in. diameter pipe if it delivers 740 gpm?

First, convert the number of gallons per minute to cubic feet per second.

$$\text{Number of ft}^3\text{/s} = \frac{740 \text{ gpm}}{(7.48 \text{ gal/ft}^3)(60 \text{ s/min})} = 1.65 \text{ ft}^3\text{/s}$$

(continued)

Next, convert the diameter from inches to feet.

$$\text{Number of ft} = (10.0 \text{ in.})(1 \text{ ft}/12 \text{ in.}) = 0.833$$

The equation is:

Q(Flow): V(Velocity)A(Area) or Flow, ft^3/s = (Area, ft^2)(Velocity, ft/s)

where the area = $(0.785)(\text{Diameter})^2$

$$1.65 \text{ ft}^3/\text{s} = (0.785)(0.833)(0.833)(\text{Flow, ft/s})$$

Rearrange and solve for the flow in feet per second.

$$\text{Flow, ft/s} = \frac{1.65 \text{ ft}^3/\text{s}}{(0.785)(0.833 \text{ ft})(0.833 \text{ ft})} = 3.029 \text{ ft/s, round to } \mathbf{3.0 \text{ ft/s}}$$

67. What is the velocity of flow in feet per second for a 6-in. diameter pipe if it delivers 350 gpm?

First, convert the number of gallons per minute to cubic feet per second.

$$\text{Number of ft}^3/\text{s} = \frac{350 \text{ gpm}}{(7.48 \text{ gal/ft}^3)(60 \text{ s/min})} = 0.78 \text{ ft}^3/\text{s}$$

Next, convert the diameter from inches to feet.

$$\text{Number of ft} = (6 \text{ in.})(1 \text{ ft}/12 \text{ in.}) = 0.5$$

The equation is:

$$\text{Flow, ft}^3/\text{s} = (\text{Area, ft}^2)(\text{Velocity, ft/s})$$

where the area = $(0.785)(\text{Diameter})^2$

$$0.78 \text{ ft}^3/\text{s} = (0.785)(0.5 \text{ ft})(0.5 \text{ ft})(\text{Flow, ft/s})$$

Rearrange and solve for the flow in feet per second.

$$\text{Flow, ft/s} = \frac{0.78 \text{ ft}^3/\text{s}}{(0.785)(0.5 \text{ ft})(0.5 \text{ ft})} = 3.97 \text{ ft/s, round to } \mathbf{4.0 \text{ ft/s}}$$

68. Water is flowing in a pipeline at 2.65 ft³/s. What is the flow in gallons per minute?

The equation is:

$$\textbf{Flow, gpm} = \textbf{(Flow, ft}^3\textbf{/s)(7.48 gal/ft}^3\textbf{)(60 s/min)}$$

Substituting:

$$\text{Flow, gpm} = (2.65\ \text{ft}^3/\text{s})(7.48\ \text{gal/ft}^3)(60\ \text{s/min}) = 1{,}189\ \text{gpm, round to } \mathbf{1{,}190\ gpm}$$

69. A water hydrant is flowing at 285 gpm. How many cubic feet per second is this?

The equation is:

$$\textbf{Flow, ft}^3\textbf{/s} = \frac{\textbf{Flow, gpm}}{\textbf{(7.48 gal/ft}^3\textbf{)(60 s/min)}}$$

Substituting:

$$\text{Flow, ft}^3/\text{s} = \frac{285\ \text{gpm}}{(7.48\ \text{gal/ft}^3)(60\ \text{s/min})} = \mathbf{0.635\ ft^3/s}$$

70. A meter indicates water is flowing from a fire hydrant at 1.50 ft³/min. How many gallons will flow from the hydrant if it is flushed for exactly 30 min?

First, convert the flow in cubic feet per minute to gallons per minute.

$$\text{Gallons per minute} = (1.50\ \text{ft}^3/\text{min})(7.48\ \text{gal/ft}^3) = 11.22\ \text{gpm}$$

Then, determine the number of gallons that flowed through the fire hydrant.

$$\text{Gallons} = (11.22\ \text{gpm})(30\ \text{min}) = 336.6\ \text{gal, round to } \mathbf{337\ gal}$$

71. A water tank with a capacity of 3.00 mil gal is being filled at a rate of 2,450 gpm. How many hours will it take to fill the tank?

First, convert gallons per minute to gallons per hour.

$$(2{,}450 \text{ gpm})(60 \text{ min/hr}) = 147{,}000 \text{ gph}$$

Next, convert million gallons to gallons.

$$\text{Number of gal} = (3.00 \text{ mil gal})(1{,}000{,}000/1 \text{ M}) = 3{,}000{,}000 \text{ gal}$$

The equation is:

$$\textbf{Time, min} = \textbf{Number of gallons/gph}$$

$$\text{Time, hr} = 3{,}000{,}000 \text{ gal}/147{,}000 \text{ gph} = 20.408 \text{ hr, round to } \mathbf{20.4 \text{ hr}}$$

DETENTION TIME CALCULATIONS

72. If a water tank has a stable volume of 2.00 mil gal and the flow into and from the tank is 4.25 mgd, what is the detention time in hours assuming no short circuiting?

The equation is:

$$\textbf{Detention time, hr} = \frac{\textbf{(Tank volume)(24 hr/d)}}{\textbf{Flow, gpd}}$$

Substituting:

First, convert million gallons to gallons.

$$(2.00 \text{ mil gal})(1{,}000{,}000/1 \text{ M}) = 2{,}000{,}000 \text{ gal}$$

And:

$$(4.25 \text{ mgd})(1{,}000{,}000/1 \text{ M}) = 4{,}250{,}000 \text{ gpd}$$

$$\text{Detention time, hr} = \frac{(2{,}000{,}000 \text{ gal})(24 \text{ hr/d})}{4{,}250{,}000 \text{ gpd}} = 11.29 \text{ hr, round to } \mathbf{11.3 \text{ hr}}$$

73. A tank averages 1,988,000 gal of water at any given time throughout the summer. If the detention time is 8.25 hr, what is the flow in million gallons per day from the tank?

First, determine the flow in gpd.

$$\textbf{Detention time, hr} = \frac{\textbf{(Tank volume)(24 hr/d)}}{\textbf{Flow, gpd}}$$

Rearrange the equation and substitute known values.

$$\text{Flow, gpd} = \frac{(1{,}988{,}000 \text{ gal})(24 \text{ hr/d})}{(8.25 \text{ hr})} = 5{,}783{,}273 \text{ gal/d}$$

Then, convert gallons per day to million gallons per day.

$$\text{Number, mgd} = \frac{5{,}783{,}273 \text{ gal/d}}{1{,}000{,}000 \text{ gal/1 M}} = \mathbf{5.78 \text{ mgd}}$$

74. Determine the detention time in hours for the following system:

Distribution pipe from the water plant to the storage tank is 1,875 ft in length and 12 in. in diameter.

The storage tank averages 1,690,000 gal of water at any given time.

The flow through the system is 4.33 mgd.

First, convert the pipe diameter in inches to feet.

$$\text{Number of ft} = (12 \text{ in.})(1 \text{ ft/12 in.}) = 1.0 \text{ ft}$$

Next, find the number of gallons in the pipeline.

$$\textbf{Number of gal} = \textbf{(0.785)(Diameter)}^2\textbf{(Length, ft)(7.48 gal/ft}^3\textbf{)}$$

$$\text{Number of gal} = (0.785)(1.0 \text{ ft})(1.0 \text{ ft})(1{,}875 \text{ ft})(7.48 \text{ gal/ft}^3) = 11{,}010 \text{ gal}$$

Add the pipe and tank volume to get the total number of gallons.

$$\text{Total number of gal} = 11{,}010 \text{ gal} + 1{,}690{,}000 \text{ gal} = 1{,}701{,}010 \text{ gal}$$

Then, convert million gallons per day to gallons per day.

$$\text{Number of gal} = (4.33 \text{ mgd})(1{,}000{,}000 \text{ gal/1 M}) = 4{,}330{,}000 \text{ gpd}$$

(continued)

Using the following equation, solve for the detention time.

The equation is:

$$\textbf{Detention time, hr} = \frac{\textbf{(Tank volume)(24 hr/d)}}{\textbf{Flow, gpd}}$$

Substituting:

$$\text{Detention time, hr} = \frac{(1{,}701{,}010 \text{ gal})(24 \text{ hr/d})}{4{,}330{,}000 \text{ gpd}} = 9.43 \text{ hr, round to } \mathbf{9.4 \text{ hr}}$$

FLUSHING TIME PROBLEMS

75. A 6.0-in. pipeline needs to be flushed. If the desired length of pipeline to be flushed is 225 ft, how many minutes will it take to flush the line at 25 gpm?

First, convert the pipe diameter from inches to feet.

$$\text{Number of ft} = (6.0 \text{ in.})(1 \text{ ft}/12 \text{ in.}) = 0.50 \text{ ft}$$

Next, determine the volume in gallons for 225 ft of the 6-in. pipeline.

The equation is:

$$\textbf{Volume, gal} = \textbf{(0.785)(Diameter)}^{\mathbf{2}}\textbf{(Length)(7.48 gal/ft}^{\mathbf{3}}\textbf{)}$$

$$\text{Volume, gal} = (0.785)(0.50)(0.50)(225 \text{ ft})(7.48 \text{ gal/ft}^3) = 330.29 \text{ gal}$$

Then, find the flushing time.

$$\text{Flushing time, min} = \frac{330.29 \text{ gal}}{25 \text{ gpm}} = 13.2 \text{ min, round to } \mathbf{13 \text{ min}}$$

76. A 28.25-ft, ¾-in. service line requires flushing. How many minutes are required to flush the service line by 25 volumes if the line is flushed at a rate of 10.0 gpm?

First, convert the pipe diameter from inches to feet.

$$\text{Number of ft} = (\tfrac{3}{4}\text{ in.})(1\text{ ft}/12\text{ in.}) = (0.75\text{ in.})(1\text{ ft}/12\text{ in.}) = 0.0625\text{ ft}$$

Next, determine the volume in gallons for 28.25 ft of the ¾-in. pipeline.

The equation is:

$$\textbf{Volume, gal} = \textbf{(0.785)(Diameter)}^2\textbf{(Length)(7.48 gal/ft}^3\textbf{)(Number of volumes)}$$

$$\text{Volume, gal} = (0.785)(0.0625)(0.0625)(28.25\text{ ft})(7.48\text{ gal/ft}^3)(25\text{ volumes}) = 16.2\text{ gal}$$

Then, find the flushing time.

$$\text{Flushing time, min} = \frac{16.20\text{ gal}}{10.0\text{ gpm}} = 1.62\text{ min, round to } \mathbf{1.6\text{ min}}$$

77. An 8.0-in. main line needs to be flushed. If the desired length of pipeline to be flushed is 810 ft, how many minutes will it take to flush the line at 25 gpm?

First, convert the pipe diameter from inches to feet.

$$\text{Number of ft} = (8.0\text{ in.})(1\text{ ft}/12\text{ in.}) = 0.667\text{ ft}$$

Next, determine the volume in gallons for 250 ft of the 8.0-in. pipeline.

The equation is:

$$\textbf{Volume, gal} = \textbf{(0.785)(Diameter)}^2\textbf{(Length)(7.48 gal/ft}^3\textbf{)}$$

$$\text{Volume, gal} = (0.785)(0.667\text{ ft})(0.667\text{ ft})(810\text{ ft})(7.48\text{ gal/ft}^3) = 2{,}115.96\text{ gal}$$

Then, find the flushing time.

$$\text{Flushing time, min} = \frac{2{,}115.96\text{ gal}}{25\text{ gpm}} = 84.64\text{ min, round to } \mathbf{85\text{ min}}$$

DENSITY PROBLEMS

Density is defined as the mass per unit volume of a substance under specific pressure and temperature conditions.

78. A substance weighs 291 g and occupies a space of 33 cm³. What is its density in grams per cubic centimeter?

The equation is:

Density = mass/volume

$$\text{Density} = 291 \text{ g}/33 \text{ cm}^3 = \mathbf{8.8 \text{ g/cm}^3}$$

79. The density of an unknown substance is 2.97 g/cm³. How much space would this substance occupy in cubic centimeters if it weighed 12.50 lb?

First, convert the number of pounds to grams.

$$\text{Number of g} = (\text{Number of lb})(454 \text{ g}/1 \text{ lb})$$

Substituting:

$$\text{Number of g} = (12.50 \text{ lb})(454 \text{ g}/1 \text{ lb}) = 5{,}675 \text{ g}$$

We know that 2.97 g of the substance occupies 1 cm^3 by knowing its density. To get the space 5,675 g occupies, divide by the density.

$$\text{Space occupied by substance} = \frac{5{,}675 \text{ g}}{2.97 \text{ g/cm}^3} = 1{,}910.77 \text{ cm}^3\text{, round to } \mathbf{1{,}910 \text{ cm}^3}$$

80. What is the density of a substance in pounds per cubic foot if it weighs 29.27 kg and occupies a space of 0.985 ft³?

The following applies:

$$1 \text{ kg} = 2.2046 \text{ lb}$$

First, convert the number of grams to pounds.

$$\text{Number of lb} = (29.27 \text{ kg})(2.2046 \text{ lb}/1 \text{ kg}) = 64.53 \text{ lb}$$

Next, divide the number of cubic feet into the number of pounds to get the density.

$$\text{Density, lb/ft}^3 = \frac{64.53 \text{ lb}}{0.985 \text{ ft}^3} = \mathbf{65.5\ lb/ft^3}$$

81. Determine the density of a substance that is 29.23 kg/ft³.

The following applies:

$$1 \text{ ft}^3 = 28{,}317 \text{ cm}^3$$

First, convert kilograms per cubic foot to grams per cubic foot.

The following applies:

$$1 \text{ kg} = 1{,}000 \text{ g}$$

$$\text{Density} = (29.23 \text{ kg/ft}^3)(1{,}000 \text{ g/kg}) = 29{,}230 \text{ g/ft}^3$$

Next, convert cubic feet by dividing by the conversion factor of 28,317 cm^3/ft^3.

$$\text{Density} = \frac{29{,}230 \text{ g/ft}^3}{28{,}317 \text{ cm}^3/\text{ft}^3} = \mathbf{1.032\ g/cm^3}$$

82. What is the density of a substance that is 32,090 g/ft³?

The following applies:

$$1 \text{ ft}^3 = 28{,}317 \text{ cm}^3$$

To find the density, divide by the conversion factor of 28,317 cm³/ft³.

$$\text{Density} = \frac{32{,}090 \text{ g/ft}^3}{28{,}317 \text{ cm}^3/\text{ft}^3} = \mathbf{1.133\ g/cm^3}$$

SPECIFIC GRAVITY OF LIQUIDS

Specific gravity (sp gr) is a dimensionless number that compares the density of one substance to that of another. Water is the standard for liquids. By definition, 1 g of water occupies a space of 1 cm³ at 4°C.

83. The density of an unknown liquid is 74.1 lb/ft³. What is the specific gravity (sp gr) of the liquid?

Water has a density of 62.4 lb/ft³. Divide the density of the unknown by the density of water.

The equation is:

$$\textbf{Specific gravity} = \textbf{Density of substance/Density of water}$$

$$\text{Specific gravity of unknown substance} = \frac{74.1 \text{ lb/ft}^3}{62.4 \text{ lb/ft}^3} = \mathbf{1.19\ sp\ gr}$$

Note: Density units are usually expressed as g/cm³ for small quantities or kg/m³ for large quantities. They can also be expressed as lb/ft³, as in this problem, or as lb/gal.

84. What is the specific gravity for a solution that weighs 9.44 lb/gal?

The density of water can also be expressed as pounds per gallon, or 8.34 lb/gal.

$$\text{Specific gravity} = \frac{9.44 \text{ lb/gal}}{8.34 \text{ lb/gal}} = \mathbf{1.13\ sp\ gr}$$

METER ACCURACY CALCULATIONS

85. A meter being tested by a laboratory shows a reading of 1,023 gal. A volumetric tank used to measure the water that flowed through the meter indicates the actual volume is 1,044 gal. What is the percent accuracy of the meter to the nearest 100th percent?

The equation is:

$$\textbf{Meter accuracy, \%} = \frac{\textbf{(Meter reading, gal)(100\%)}}{\textbf{Actual volume, gal}}$$

$$\text{Meter accuracy, \%} = \frac{(1{,}023 \text{ gal})(100\%)}{1{,}044 \text{ gal}} = \textbf{97.99\% Meter accuracy}$$

86. A meter being tested reads 1,288 L. A volumetric tank shows the actual value is 1,321 L. What is the percent accuracy of the meter to the nearest 100th percent?

The equation is:

$$\textbf{Meter accuracy, \%} = \frac{\textbf{(Meter reading, gal)(100\%)}}{\textbf{Actual volume, gal}}$$

$$\text{Meter accuracy, \%} = \frac{(1{,}288 \text{ L})(100\%)}{1{,}321 \text{ L}} = \textbf{97.50\% Meter accuracy}$$

CALCULATIONS FOR CHLORINE DEMAND AND RESIDUAL

87. A water treatment plant injects chlorine at a dosage rate of 3.00 mg/L after the filters. The chlorine residual is 1.15 mg/L at a distant point in the distribution system. What must have been the chlorine demand between the two sampling points?

Write the equation to solve for the unknown chlorine demand.

Chlorine demand, mg/L = Chlorine dose, mg/L – Chlorine residual, mg/L

Chlorine demand, mg/L = 3.00 mg/L – 1.15 mg/L = **1.85 mg/L**

88. A treatment plant produces 8.2 mgd with a chlorine dosage of 3.15 mg/L. If a distant storage tank has a chlorine residual of 1.83, what has been the chlorine demand of the distribution system between the water plant and the storage tank?

The equation is:

Chlorine demand, mg/L = Chlorine dose, mg/L – Chlorine residual, mg/L

Chlorine demand, mg/L = 3.15 mg/L – 1.83 mg/L = **1.32 mg/L**

89. What is the chlorine residual in a system that has a chlorine dosage of 2.75 mg/L and a chlorine demand of 1.93 mg/L?

The equation is:

Chlorine residual, mg/L = Chlorine dose, mg/L – Chlorine demand, mg/L

Chlorine residual, mg/L = 2.75 mg/L – 1.93 mg/L = **0.82 mg/L**

DOSAGE CALCULATIONS

Note: For convenience, most of the dosage problems in this book do not take into account a chlorine demand.

90. How many pounds per day of chlorine gas are required to treat 9.23 mgd if the dosage is 2.50 mg/L?

The equation is:

Pounds per day = (mgd)(Dosage, mg/L)(8.34 lb/d)

Chlorine, lb/d = (9.23 mgd)(2.50 mg/L)(8.34 lb/gal)

= 192.44 lb/d, round to **192 lb/d of Cl_2**

91. A water treatment plant is feeding an average of 345 lb/d of chlorine. If the dosage is 2.50 mg/L, what is the number of million gallons per day being treated?

The equation is:

$$\textbf{Pounds per day} = \textbf{(mgd)(Dosage, mg/L)(8.34 lb/d)}$$

Rearrange to solve for million gallons per day.

$$\text{Number of mgd} = \frac{\text{Number of lb/d of chlorine}}{\text{(Dosage, mg/L)(8.34 lb/gal)}}$$

$$\text{Number of mgd} = \frac{\text{345 lb/d}}{\text{(2.50 mg/L)(8.34 lb/gal)}} = \textbf{16.5 mgd}$$

92. What is the dosage in milligrams per liter for a treatment plant that injects 575 lb/d of chlorine and treats 20.5 mgd?

The equation is:

$$\textbf{Pounds per day} = \textbf{(mgd)(Dosage, mg/L)(8.34 lb/d)}$$

Rearrange to solve for dosage.

$$\text{Dosage, mg/L} = \frac{\text{Number of lb/d of chlorine}}{\text{(mgd)(8.34 lb/gal)}}$$

$$\text{Dosage, mg/L} = \frac{\text{575 lb/d}}{\text{(20.5 mgd)(8.34 lb/gal)}} = \textbf{3.36 mg/L}$$

93. How many pounds per day of chlorine are needed to treat 16.52 mgd if the dosage is 3.10 mg/L?

The equation is:

$$\textbf{Pounds per day} = \textbf{(mgd)(Dosage, mg/L)(8.34 lb/d)}$$

$$\begin{aligned}\text{Chlorine, lb/day} &= \text{(16.52 mgd)(3.10 mg/L)(8.34 lb/gal)}\\ &= \text{427.1 lb/d, round to } \textbf{427 lb/d of } \mathbf{Cl_2}\end{aligned}$$

94. What is the dosage in milligrams per liter for a treatment plant that uses 855 lb/d of chlorine and treats 45.25 mgd?

The equation is:

Pounds per day = (mgd)(Dosage, mg/L)(8.34 lb/d)

Rearrange to solve for dosage.

$$\text{Dosage, mg/L} = \frac{\text{Number of lb/d of Chlorine}}{\text{(mgd)(8.34 lb/gal)}}$$

$$\text{Dosage, mg/L} = \frac{\text{855 lb/d}}{\text{(45.25 mgd)(8.34 lb/gal)}} = \text{2.266 mg/L, round to } \textbf{2.27 mg/L}$$

95. A water treatment plant is feeding an average of 210 lb/d of chlorine. If the dosage is 3.25 mg/L, what is the number of million gallons per day being treated?

The equation is:

Pounds per day = (mgd)(Dosage, mg/L)(8.34 lb/d)

Rearrange to solve for million gallons per day.

$$\text{Number of mgd} = \frac{\text{Number of lb/d of chlorine}}{\text{(Dosage, mg/L)(8.34 lb/gal)}}$$

$$\text{Number of mgd} = \frac{\text{210 lb/d}}{\text{(3.25 mg/L)(8.34 lb/gal)}} = \text{7.748 mgd, round to } \textbf{7.7 mgd}$$

96. What should the setting be on a chlorinator in pounds per day if the dosage desired is 2.70 mg/L and the pumping rate from the well is 845 gpm?

First, convert the pumping rate to million gallons per day.

The equation is:

$$\textbf{Million gallons per day} = \frac{\textbf{(Pumping rate, gpm)(1,440 min/d)}}{\textbf{1,000,000/M}}$$

Substituting:

$$\text{Million gallons per day} = \frac{(845\ \text{gpm})(1{,}440\ \text{min/d})}{1{,}000{,}000/1\ \text{M}} = 1.217\ \text{mgd}$$

Next, use the "pounds" equation to solve the problem.

The equation is:

$$\textbf{Chlorine, lb/d} = \textbf{(mgd)(Dosage, mg/L)(8.34 lb/gal)}$$

$$\text{Chlorine, lb/d} = (1.217\ \text{mgd})(2.70\ \text{mg/L})(8.34\ \text{lb/gal})$$

$$= 27.40\ \text{lb/d, round to } \mathbf{27.4\ lb/d\ of\ Cl_2}$$

97. A small tank containing 1,200 gal of water needs to be disinfected to be put back into service. If the dosage needed is 50 mg/L, how many pounds of calcium hypochlorite (60.5% available chlorine) are required?

First, convert gallons to million gallons.

$$\text{Number of mil gal} = (1{,}200\ \text{gal})(1\ \text{M}/1{,}000{,}000) = 0.0012\ \text{mil gal}$$

Next, determine the pounds of chlorine needed.

The equation is:

$$\textbf{Number of lb, } \mathbf{Cl_2} = \textbf{(Million gallons)(Dosage, mg/L)(8.34 lb/gal)}$$

$$\text{Number of lb, } Cl_2 = (0.0012\ \text{mil gal})(50\ \text{mg/L})(8.34\ \text{lb/gal}) = 0.5\ \text{lb of } Cl_2$$

Next, find the number of pounds of calcium hypochlorite:

$$\frac{\text{Number of lb, } Cl_2}{\text{Number of lb, } Ca(OCl)_2} = \frac{\text{Percent } Cl_2 \text{ in } Ca(OCl)_2}{\text{Percent } Ca(OCl)_2}$$

$$\frac{0.5\ \text{lb of } Cl_2}{\text{Number of lb, } Ca(OCl)_2} = \frac{60.5\%\ Cl_2 \text{ in } Ca(OCl)_2}{100\%\ Ca(OCl)_2}$$

(continued)

$$60.5\%\ Cl_2 \text{ in } Ca(OCl)_2[\text{Number of lb, } Ca(OCl)_2] = 0.5 \text{ lb of } Cl_2[100\%\ Ca(OCl)_2]$$

$$\text{Number of lb, } Ca(OCl)_2 = \frac{0.5 \text{ lb of } Cl_2[100\%\ Ca(OCl)_2]}{60.5\%\ Cl_2 \text{ in } Ca(OCl)_2}$$

$$\text{Number of lb, } Ca(OCl)_2 = \mathbf{0.8\ lb\ of\ Ca(OCl)_2}$$

98. How many gallons of a sodium hypochlorite solution that contains 5.50% available chlorine are needed to disinfect a 1.0-ft diameter pipeline that is 655 ft long if the dosage required is 25.0 mg/L? Assume the sodium hypochlorite is 8.95 lb/gal.

First, find the volume of the pipe.

$$\mathbf{Volume,\ gal = (0.785)(Diameter)^2(Length)(7.48\ gal/ft^3)}$$

$$\text{Volume} = (0.785)(1.0 \text{ ft})(1.0 \text{ ft})(655 \text{ ft})(7.48 \text{ gal/ft}^3) = 3{,}846 \text{ gal}$$

Next, find the number of million gallons.

$$\text{Million gallons} = (3{,}846 \text{ gal})(1 \text{ M}/1{,}000{,}000) = 0.003846 \text{ mil gal}$$

Then, use the "pounds" equation:

$$\mathbf{Pounds\ of\ sodium\ hypochlorite\ solution = \frac{(mil\ gal)(Dosage,\ mg/L)(8.34\ lb/gal)}{Percent\ available\ chlorine/100\%}}$$

$$\text{Pounds of sodium hypochlorite solution} = \frac{(0.003846 \text{ mil gal})(25.0 \text{ mg/L})(8.34 \text{ lb/gal})}{5.50\% \text{ available chlorine}/100\%} = 14.58 \text{ lb}$$

Last, calculate the number of gallons of sodium hypochlorite.

$$\text{Sodium hypochlorite, gal} = 14.58 \text{ lb}/8.95 \text{ lb/gal}$$

$$= 1.63 \text{ gal, round to } \mathbf{1.6\ gal\ of\ sodium\ hypochlorite}$$

99. A treatment plant uses 340 lb/d of chlorine gas. If the chlorine demand is 1.95 mg/L and the chlorine residual is 1.10 mg/L, how many million gallons per day are being treated?

First, find the total chlorine dosage.

The equation is:

$$\textbf{Total } \mathbf{Cl_2} \textbf{ dosage} = \mathbf{Cl_2} \textbf{ demand} + \mathbf{Cl_2} \textbf{ residual}$$

$$\text{Total } Cl_2 \text{ dosage} = 1.95 \text{ mg/L (Demand)} + 1.10 \text{ mg/L (Residual)} = 3.05 \text{ mg/L}$$

Then, use the "pounds" equation, but solve for the unknown, million gallons per day.

$$\textbf{Pounds per day} = \textbf{(mgd)(Dosage, mg/L)(8.34 lb/gal)}$$

$$\text{Million gallons per day} = \frac{\text{lb/d}}{\text{(Dosage, mg/L)(8.34 lb/gal)}}$$

$$\text{Million gallons per day} = \frac{340 \text{ lb/d}}{(3.05 \text{ mg/L})(8.34 \text{ lb/gal})} = 13.37 \text{ mgd, round to } \mathbf{13 \text{ mgd}}$$

100. How many gallons of a sodium hypochlorite solution that contains 5.25% available chlorine are needed to disinfect a water main that is 24 in. in diameter and a pipeline that is 716 ft long if the dosage required is 25.0 mg/L? Assume the sodium hypochlorite solution weighs 8.94 lb/gal.

First, convert inches to feet.

$$24 \text{ in./12 in. per foot} = 2.0 \text{ ft}$$

Next, find the volume of the pipe.

$$\textbf{Volume, gal} = \mathbf{(0.785)(Diameter)^2(Length)(7.48\ gal/ft^3)}$$

$$\text{Volume} = (0.785)(2.0 \text{ ft})(2.0 \text{ ft})(716 \text{ ft})(7.48 \text{ gal/ft}^3) = 16{,}817 \text{ gal}$$

Next, find the number of million gallons.

$$\text{Million gallons} = (16{,}817 \text{ gal})(1 \text{ M}/1{,}000{,}000) = 0.0168 \text{ mil gal}$$

(continued)

Then, use the "pounds" equation.

$$\textbf{Pounds of sodium hypochlorite} = \frac{\textbf{(mil gal)(Dosage, mg/L)(8.34 lb/gal)}}{\textbf{Percent available chlorine/100\%}}$$

$$\text{Pounds of sodium hypochlorite} = \frac{(0.0168 \text{ mil gal})(25.0 \text{ mg/L})(8.34 \text{ lb/gal})}{5.25\% \text{ available chlorine}/100\%} = 66.72 \text{ lb}$$

Last, calculate the number of gallons of sodium hypochlorite.

Sodium hypochlorite, gal = 66.72 lb/8.94 lb/gal = **7.5 gal of sodium hypochlorite**

101. A 4.0-mil gal storage tank needs to be disinfected with a calcium hypochlorite tablet that contains 60.5% available chlorine. The tank is to be filled at exactly 10% capacity and the initial chlorine dosage desired is 25.0 mg/L. How many pounds of calcium hypochlorite are required?

First, find the initial amount of water to be disinfected.

$$10\% \text{ capacity} = (4.0 \text{ mil gal})(10\%/100\%) = 0.40 \text{ mil gal}$$

Next, determine the number of pounds of chlorine needed by using the "pounds" equation.

$$\textbf{Number of lb} = \frac{\textbf{(mil gal)(Dosage, mg/L)(8.34 lb/gal)}}{\textbf{Percent available chlorine/100\%}}$$

$$Ca(OCl)_2\text{, lb} = \frac{(0.40 \text{ mil gal})(25.0 \text{ mg/L})(8.34 \text{ lb/gal})}{60.5\%/100\%}$$

= 137.8 lb, round to **140 lb of** $\mathbf{Ca(OCl)_2}$

102. A 1.0-mil gal storage tank needs to be disinfected with a calcium hypochlorite solution that has 60.2% available chlorine. The tank is to be filled to exactly 10% capacity and the initial chlorine dosage required is 25 mg/L. How many pounds of calcium hypochlorite are needed?

First, find the initial amount of water to be disinfected.

$$10\% \text{ capacity} = (1.0 \text{ mil gal})(10\%/100\%) = 0.10 \text{ mil gal}$$

Next, determine the number of pounds of chlorine needed by using the "pounds" equation.

The equation is:

$$\textbf{Number of lb} = \frac{\textbf{(mil gal)(Dosage, mg/L)(8.34 lb/gal)}}{\textbf{Percent available chlorine/100\%}}$$

$$\text{Calcium hypochlorite, lb} = \frac{(0.10 \text{ mil gal})(25 \text{ mg/L})(8.34 \text{ lb/gal})}{60.2\%/100\%}$$

= 34.63 lb, round to **35 lb of calcium hypochlorite**

103. A 3.0-mil gal storage tank needs to be disinfected with a calcium hypochlorite solution that contains 59.5% available chlorine. The tank is to be filled to exactly 10% capacity and the initial chlorine dosage required is 20 mg/L. How many pounds of calcium hypochlorite are needed?

First, find the initial amount of water to be disinfected.

$$10\% \text{ capacity} = (3.0 \text{ mil gal})(10\%/100\%) = 0.30 \text{ mil gal}$$

Determine the number of pounds of chlorine needed by using the "pounds" equation.

$$\textbf{Number of lb} = \frac{\textbf{(mil gal)(Dosage, mg/L)(8.34 lb/gal)}}{\textbf{Percent available chlorine/100\%}}$$

$$\text{Calcium hypochlorite, lb} = \frac{(0.30 \text{ mil gal})(20.0 \text{ mg/L})(8.34 \text{ lb/gal})}{59.5\%/100\%}$$

= 84.1 lb, round to **84 lb of calcium hypochlorite**

104. How many gallons of a sodium hypochlorite solution that contains 9.6% available chlorine are needed to disinfect a water main that is 18 in. in diameter and a pipeline that is 1,245 ft long if the dosage required is 25.0 mg/L? Assume the sodium hypochlorite solution weighs 9.45 lb/gal.

First, convert 18 in. to feet.

18 in./12 in. per foot = 1.5 ft

Next, find the volume of the pipe.

Volume, gal = (0.785)(Diameter)^2(Length)(7.48 gal/ft^3)

Volume = (0.785)(1.5 ft)(1.5 ft)(1,245 ft)(7.48 gal/ft^3) = 16,448 gal

Next, find the number of million gallons.

Million gallons = (16,448 gal)(1 M/1,000,000) = 0.016448 mil gal

Then, use the "pounds" equation:

$$\textbf{Pounds of sodium hypochlorite} = \frac{\textbf{(mil gal)(Dosage, mg/L)(8.34 lb/gal)}}{\textbf{Percent available chlorine/100\%}}$$

$$\text{Pounds of sodium hypochlorite} = \frac{(0.016448 \text{ mil gal})(25.0 \text{ mg/L})(8.34 \text{ lb/gal})}{9.6\% \text{ available chlorine}/100\%} = 35.723 \text{ lb}$$

Last, calculate the number of gallons of sodium hypochlorite.

Sodium hypochlorite, gal = 35.723 lb/9.45 lb/gal

= 3.78 gal, round to **3.8 gal of sodium hypochlorite**

105. Determine the number of gallons of a sodium hypochlorite solution that contains 9.0% available chlorine needed to disinfect a new water main that is 24.0 in. in diameter and a pipeline that is 2,300 ft long if the dosage required is 50.0 mg/L. Assume the sodium hypochlorite solution weighs 9.36 lb/gal.

First, convert 24.0 in. to feet.

24.0 in./12 in. per foot = 2.00 ft

Next, find the volume of the pipe.

Volume, gal = (0.785)(Diameter)^2(Length)(7.48 gal/ft^3)

Volume = (0.785)(2.00 ft)(2.00 ft)(2,300 ft)(7.48 gal/ft^3) = 54,021 gal

Next, find the number of million gallons.

Million gallons = (54,021 gal)(1 M/1,000,000) = 0.054021 mil gal

Then, use the "pounds" equation:

$$\textbf{Pounds of sodium hypochlorite} = \frac{\textbf{(mil gal)(Dosage, mg/L)(8.34 lb/gal)}}{\textbf{Percent available chlorine/100\%}}$$

$$\text{Pounds of sodium hypochlorite} = \frac{(0.054021 \text{ mil gal})(50.0 \text{ mg/L})(8.34 \text{ lb/gal})}{9.0\% \text{ available chlorine}/100\%} = \textbf{250.3 lb}$$

Last, calculate the number of gallons of sodium hypochlorite.

Sodium hypochlorite, gal = 250.3 lb/9.36 lb/gal

= 26.74 gal, round to **27 gal of sodium hypochlorite**

Distribution Grade 1 Test

(answers on p. 46)

1. Determine the daily average total trihalomethanes (TTHM) given the following data:

Monday	0.0073 mg/L
Tuesday	0.0085 mg/L
Wednesday	0.0094 mg/L
Thursday	0.0079 mg/L
Friday	0.0092 mg/L

2. What is the difference in temperature in degrees Fahrenheit between the water source and the processed water in a storage tank if the temperature of the water source is 11.4°C and the water in the storage tank is 14.8°C?

3. Find the area of a tank that is 148 ft in diameter.

4. What is the pounds per square inch at the bottom of a tank if the water level is 38.29 ft deep?

5. If a trench is 526 ft long, 4.0 ft wide, and 5.5 ft deep, how many cubic yards of soil were excavated?

6. If exactly 100 gal of polymer costs $19.50, what will 5,500 gal cost, assuming no quantity discount?

7. What is the velocity of flow in feet per second for an 8.0-in. diameter pipe if it delivers 675 gpm?

8. What should the setting be on a chlorinator in pounds per day if the dosage desired is 2.90 mg/L and the pumping rate from the well is 975 gpm?

9. A treatment plant uses 278 lb/d of chlorine gas. If the chlorine demand is 0.85 mg/L and the chlorine residual is 1.50 mg/L, how many million gallons per day are being treated?

10. A water tank that is 105 ft in diameter needs to be disinfected with a 5.0% sodium hypochlorite solution. If the tank is to be filled to only a depth of 5.0 ft and the concentration required is 20.0 mg/L, how many gallons of sodium hypochlorite are needed? Assume the sodium hypochlorite solution weighs 8.92 lb/gal.

Solutions to Distribution Grade 1 Test

1. The equation is:

$$\textbf{Average TTHM} = \frac{\textbf{Sum of TTHM measurements}}{\textbf{Number of measurements}}$$

Average TTHM =

$$\frac{0.0073 \text{ mg/L} + 0.0085 \text{ mg/L} + 0.0094 \text{ mg/L} + 0.0079 \text{ mg/L} + 0.0092 \text{ mg/L}}{5}$$

Average TTHM = **0.0085 mg/L**

2. First, find the difference in degrees Celsius.

°C = 14.8°C – 11.4°C = 3.4°C

The equation for Fahrenheit is:

$$\textbf{°F} = (9/5)\textbf{°C} + 32$$

°F = (9/5)3.4°C + 32 = 38.12°F, round to **38.1°F**

3. The equation is:

$$\textbf{Area} = (\pi)(r)^2$$

Radius (r) = Diameter/2 = 148 ft/2 = 74 ft

Area of tank = (3.14)(74 ft)(74 ft) =17,195 ft^2, round to **17,200 ft^2**

4. The equation is:

$$\mathbf{psi} = \frac{\mathbf{Depth, ft}}{\mathbf{2.31\ ft/psi}}$$

$$\text{psi} = \frac{38.29\ \text{ft}}{2.31\ \text{ft/psi}} = \mathbf{16.6\ psi}$$

5. First, find the number of cubic feet.

The equation is:

$$\mathbf{Volume,\ ft^3 = (Length)(Width)(Depth)}$$

$$\text{Volume, ft}^3 = (526\ \text{ft})(4.0\ \text{ft})(5.5\ \text{ft}) = 11{,}572\ \text{ft}^3$$

Next, convert cubic feet to cubic yards.

The following applies:

$$1\ \text{yd}^3 = 27\ \text{ft}^3$$

Thus,

$$\frac{11{,}572\ \text{ft}^3}{27\ \text{ft}^3/\text{yd}^3} = 428.59\ \text{yd}^3\text{, round to } \mathbf{430\ yd^3}$$

6. Solve the problem using proportions.

$$\$19.50{:}100\ \text{gal} = x{:}5{,}500\ \text{gal}$$

Put in fraction form and solve for x.

$$\frac{\$19.50}{100\ \text{gal}} = \frac{(x)}{5{,}500\ \text{gal}}$$

$$x = \frac{(\$19.50)(5{,}500\ \text{gal})}{100\ \text{gal}} = \mathbf{\$1{,}072.50}$$

7. First, convert the number of gallons per minute to cubic feet per second.

$$\text{Number of ft}^3\text{/s} = \frac{675 \text{ gpm}}{(7.48 \text{ gal/ft}^3)(60 \text{ s/min})} = 1.504 \text{ ft}^3\text{/s}$$

Next, convert the diameter from inches to feet.

$$\text{Number of ft} = (8.0 \text{ in.})(1 \text{ ft/12 in.}) = 0.667$$

The equation is:

$$\text{Flow, ft}^3\text{/s} = (\text{Area, ft}^2)(\text{Velocity, ft/s})$$

where the area = $(0.785)(\text{Diameter})^2$

$$1.504 \text{ ft}^3\text{/s} = (0.785)(0.667 \text{ ft})(0.667 \text{ ft})(\text{Velocity, ft/s})$$

Rearrange and solve for the flow in feet per second.

$$\text{Velocity, ft/s} = \frac{1.504 \text{ ft}^3\text{/s}}{(0.785)(0.667 \text{ ft})(0.667 \text{ ft})} = 4.307 \text{ ft/s, round to } \mathbf{4.3 \text{ ft/s}}$$

8. First, convert the pumping rate to million gallons per day.

The equation is:

$$\textbf{Million gallons per day} = \frac{\textbf{(Pumping rate, gpm)(1,440 min/d)}}{\textbf{1,000,000/M}}$$

Substituting:

$$\text{Million gallons per day} = \frac{(975 \text{ gpm})(1{,}440 \text{ min/d})}{1{,}000{,}000/1 \text{ M}} = 1.404 \text{ mgd}$$

Next, use the "pounds" equation to solve the problem.

The equation is:

$$\textbf{Chlorine, lb/d} = \textbf{(mgd)(Dosage, mg/L)(8.34 lb/gal)}$$

$$\text{Chlorine, lb/d} = (1.404 \text{ mgd})(2.90 \text{ mg/L})(8.34 \text{ lb/gal})$$

$$= 33.96 \text{ lb/d, round to } \mathbf{34 \text{ lb/d of } Cl_2}$$

9. First, find the total chlorine dosage.

The equation is:

$$\textbf{Total Cl}_2\textbf{ dosage} = \textbf{Cl}_2\textbf{ demand} + \textbf{Cl}_2\textbf{ residual}$$

$$\text{Total Cl}_2 \text{ dosage} = 0.85 \text{ mg/L (Demand)} + 1.50 \text{ mg/L (Residual)} = 2.35 \text{ mg/L}$$

Then, use the "pounds" equation, but solve for the unknown, million gallons per day.

$$\textbf{Pounds per day} = \textbf{(mgd)(Dosage, mg/L)(8.34 lb/gal)}$$

$$\text{Million gallons per day} = \frac{\text{lb/d}}{\text{(Dosage)(8.34 lb/gal)}}$$

$$\text{Million gallons per day} = \frac{278 \text{ lb/d}}{(2.35 \text{ mg/L})(8.34 \text{ lb/gal})} = 14.18 \text{ mgd, round to } \textbf{14 mgd}$$

10. First, find the volume of the tank.

$$\textbf{Volume, gal} = \textbf{(0.785)(Diameter)}^2\textbf{(Depth)(7.48 gal/ft}^3\textbf{)}$$

$$\text{Volume} = (0.785)(105 \text{ ft})(105 \text{ ft})(5.0 \text{ ft})(7.48 \text{ gal/ft}^3) = 323{,}683 \text{ gal}$$

Next, find the number of million gallons.

$$\text{Million gallons} = (323{,}683 \text{ gal})(1 \text{ M}/1{,}000{,}000) = 0.324 \text{ mil gal}$$

Then, use the "pounds" equation:

$$\textbf{Pounds of sodium hypochlorite} = \frac{\textbf{(mil gal)(Dosage, mg/L)(8.34 lb/gal)}}{\textbf{Percent available chlorine/100\%}}$$

$$\text{Pounds of sodium hypochlorite} = \frac{(0.324 \text{ mil gal})(20.0 \text{ mg/L})(8.34 \text{ lb/gal})}{5.0\% \text{ available chlorine}/100\%} = 1{,}081 \text{ lb}$$

Last, calculate the number of gallons of sodium hypochlorite.

$$\text{Sodium hypochlorite, gal} = 1{,}081 \text{ lb}/8.92 \text{ lb/gal} = 121.19 \text{ gal}$$

Sodium hypochlorite, gal = 121.19 gal, round to **120 gal of sodium hypochlorite**

CHAPTER

2

DISTRIBUTION

Grade 2

Students preparing for the grade 3 and grade 4 distribution certification tests should also understand these problems.

CONVERSION PROBLEMS

1. **How many liters are there in exactly 44 gal?**

   There are 3.785 L per gallon, thus:

   Number of L = (44 gal)(3.785 L/gal) = 166.54 L, round to **166.5 L**

   Note: The 44 gal is not a measurement, but a specific amount asked for; therefore, it has an infinite number of significant figures. The 3.785 L determines the number of significant figures.

2. **How many gallons are there in 6,825 L?**

   There is 1 gal per 3.785 L, thus:

$$\text{Number of gal} = \frac{6{,}825 \text{ L}}{3.785 \text{ L/gal}} = \mathbf{1{,}803 \text{ gal}}$$

3. Convert 122 ft^3 to gallons and liters.

First, find the number of gallons.

$$\text{Number of gal} = (\text{Number, ft}^3)(7.48\ \text{gal/ft}^3)$$

$$\text{Number of gal} = (122\ \text{ft}^3)(7.48\ \text{gal/ft}^3) = 912.56\ \text{gal, round to } \mathbf{913\ gal}$$

Next, convert the number of gallons, 912.56, to liters.

$$\text{Number of L} = (\text{Number of gal})(3.785\ \text{L/gal})$$

$$\text{Number of L} = (912.56\ \text{gal})(3.785\ \text{L/gal}) = 3{,}454\ \text{L, round to } \mathbf{3{,}450\ L}$$

Note: We use 912.56 gal because we could have done the math operation in one step, i.e., Number of L = (Number, ft^3)(7.48 gal/ft^3)(3.785 L/gal).

4. Convert 2,336 lb of water to cubic feet.

There are 8.34 lb/gal and 7.48 gal/ft^3.

$$\text{Number of ft}^3 = \frac{2{,}336\ \text{lb}}{(8.34\ \text{lb/gal})(7.48\ \text{gal/ft}^3)} = \mathbf{37.4\ ft^3}$$

5. Convert 1.55 mgd to cubic feet per second.

The equation is:

$$\textbf{Number, ft}^3\textbf{/s} = \frac{\textbf{(Number, mgd)(1,000,000 gal)(1 ft}^3\textbf{)(1 d)(1 min)}}{\textbf{1 mil gal (7.48 gal)(1,440 min)(60 s)}}$$

or:

$$\textbf{Number, ft}^3\textbf{/s} = \frac{\textbf{(Number, mgd)(1,000,000 gal)(1 ft}^3\textbf{)(1 d)}}{\textbf{1 mil gal (7.48 gal)(86,400 s)}}$$

$$\text{Number, ft}^3\text{/s} = \frac{(1.55\ \text{mgd})(1{,}000{,}000\ \text{gal})(1\ \text{ft}^3)(1\ \text{d})(1\ \text{min})}{1\ \text{mil gal}\ (7.48\ \text{gal})(1{,}440\ \text{min})(60\ \text{s})} = \mathbf{2.40\ ft^3/s}$$

6. Convert 25 L/s to gallons per minute.

The equation is:

$$\textbf{Flow, gpm} = \frac{\textbf{(L/s)(60 s/min)}}{\textbf{3.785 L/gal}}$$

Substituting:

$$\text{Flow, gpm} = \frac{(25\text{ L/s})(60\text{ s/min})}{3.785\text{ L/gal}} = 396\text{ gpm, round to }\textbf{400 gpm}$$

7. Convert 825 gpm to liters per second.

The equation is:

$$\textbf{Flow, L/s} = \frac{\textbf{(Flow, gpm)(3.785 L/gal)}}{\textbf{60 s/min}}$$

$$\text{Flow, L/s} = \frac{(825\text{ gpm})(3.785\text{ L/gal})}{60\text{ s/min}} = \textbf{52.0 L/s}$$

8. How many gallons are there in 22,440 ft^3?

The equation is:

$$\textbf{Number of gal} = \textbf{(Number, ft}^3\textbf{)(7.48 gal/ft}^3\textbf{)}$$

$$\text{Number of gal} = (22{,}440\text{ ft}^3)(7.48\text{ gal/ft}^3) = 167{,}851.2\text{ gal, round to }\textbf{168,000 gal}$$

9. Convert 77.4 ft³/s to million gallons per day.

The equation is:

$$\textbf{Number of mgd} = \frac{(\text{Number, ft}^3)(86{,}400\text{ s})(7.48\text{ gal})\ (1\text{ mil gal})}{\text{sec}\quad \text{day}\quad \text{ft}^3\quad 1{,}000{,}000\text{ gal}}$$

$$\text{Number of mgd} = \frac{(77.4\text{ ft}^3)(86{,}400\text{ s})(7.48\text{ gal})\ (1\text{ mil gal})}{\text{sec}\quad \text{day}\quad \text{ft}^3\quad 1{,}000{,}000\text{ gal}} = \mathbf{50.00\ mgd}$$

TEMPERATURE CONVERSIONS

10. Convert 25°C to degrees Fahrenheit.

The equation for Fahrenheit is:

$$\mathbf{°F = (9/5)°C + 32}$$

$$°F = (9/5)25°C + 32 = \mathbf{77°F}$$

11. Convert 90.0°F to degrees Celcius.

$$\mathbf{°C = 5/9(°F - 32)}$$

$$°C = 5/9(90.0°F - 32) = \mathbf{32.2°C}$$

12. Convert 55°C to degrees Fahrenheit.

$$°F = (9/5)55°C + 32 = \mathbf{131°F}$$

13. Convert 32°F to degrees Celsius.

$$°C = \frac{5}{9}(32°F - 32) = \mathbf{0°C}$$

PERCENT AND PARTS PER MILLION CONVERSION PROBLEMS

14. Convert a solution that has 120,000 ppm to percent.

The equation is:

$$\textbf{Percent} = \frac{\textbf{(Number of ppm)(100\%)}}{\textbf{1,000,000/1 M}}$$

$$\text{Percent} = \frac{(120{,}000 \text{ ppm})(100\%)}{1{,}000{,}000/1 \text{ M}} = \textbf{12\% solution}$$

15. Convert 333 ppm to percent.

$$\text{Percent} = \frac{(333 \text{ ppm})(100\%)}{1{,}000{,}000/1 \text{ M}} = \textbf{0.0333\%}$$

16. Convert a solution that has 7,250 ppm to percent.

Another method to solving these types of conversion problems is as follows:

The following applies:

$$1\% \text{ solution} = 10{,}000 \text{ ppm}$$

Thus,

$$\textbf{Percent solution} = \frac{\textbf{Number of ppm}}{\textbf{10,000 ppm/1\%}}$$

$$\text{Percent solution} = \frac{7{,}250 \text{ of ppm}}{10{,}000 \text{ ppm}/1\%} = \textbf{0.725\% solution}$$

17. Convert a 9.5% solution to parts per million.

The equation is:

$$\textbf{Parts per million} = \frac{\textbf{(Percent)(1,000,000)}}{\textbf{(100\%)(1 M)}}$$

$$\text{Parts per million} = \frac{(9.5\%)(1{,}000{,}000)}{(100\%)(1\ \text{M})} = \textbf{95,000 ppm}$$

18. What is the percent chlorine in a tank that has a concentration of 1.5 mg/L?

The following applies:

$$1\ \text{mg/L} = 1\ \text{ppm}$$

Thus,

$$1.5\ \text{mg/L} = 1.5\ \text{ppm}$$

The equation is:

$$\textbf{Percent solution} = \frac{\textbf{Number of ppm}}{\textbf{10,000 ppm/1\%}}$$

$$\text{Percent chlorine} = \frac{1.5\ \text{ppm}}{10{,}000\ \text{ppm}/1\%} = \textbf{0.00015\%}$$

AVERAGE CALCULATIONS

19. What is the average iron concentration from a well given the following data (in milligrams per liter)?

1	2	3	4	5	6	7
0.37	0.52	0.35	0.41	0.43	0.41	0.38

First, add all seven measurements.

$$0.37 + 0.52 + 0.35 + 0.41 + 0.43 + 0.41 + 0.38 = 2.87\ \text{mg/L}$$

The equation is:

$$\textbf{Average} = \frac{\textbf{Sum of measurements}}{\textbf{Number of measurements}}$$

$$\text{Average, iron concentration} = \frac{2.87 \text{ mg/L}}{7} = \mathbf{0.41 \text{ mg/L}}$$

20. What is the average pH given the following data?

1	2	3	4	5	6	7	8
7.12	7.17	7.23	7.02	6.95	7.29	7.22	7.19

First, add all seven measurements.

$$7.12 + 7.17 + 7.23 + 7.02 + 6.95 + 7.29 + 7.22 + 7.19 = 57.19$$

The equation is:

$$\textbf{Average} = \frac{\textbf{Sum of measurements}}{\textbf{Number of measurements}}$$

$$\text{Average, pH} = \frac{57.19}{8} = 7.149 \text{ pH, round to } \mathbf{7.15 \text{ pH}}$$

PERCENT CALCULATIONS

21. What is 72% of 3,448?

The equation is:

$$\textbf{Number sought} = \textbf{(Percent/100\%)(Total number)}$$

$$\text{Number sought} = (72\%/100\%)(3{,}448) = 2{,}482.56, \text{ round to } \mathbf{2{,}500}$$

22. If 476 is 27.5%, what is 100%?

The equation is:

$$\textbf{100\% number} = \frac{\textbf{(Number given)(100\%)}}{\textbf{Percent of given number}}$$

$$100\% \text{ number} = \frac{(476)(100\%)}{27.5\%} = 1{,}731, \text{ round to } \mathbf{1{,}730}$$

PERCENT REMOVAL CALCULATIONS

23. What is the turbidity removal efficiency through a water plant if the source water turbidity is 22.6 ntu and the treated water entering the distribution system is 0.040 ntu? Give the answer to three significant figures.

The equation is:

$$\textbf{Percent ntu removal efficiency} = \frac{\textbf{(Influent ntu – Effluent ntu)(100\%)}}{\textbf{Influent ntu}}$$

Condensed version of above equation:

$$\text{Percent ntu removal efficiency} = \frac{(\text{In} - \text{Out})(100\%)}{\text{In}}$$

$$\text{Percent ntu removal efficiency} = \frac{(22.6 \text{ ntu} - 0.040 \text{ ntu})(100\%)}{22.6 \text{ ntu}}$$

Percent ntu removal efficiency = **99.8% ntu removal efficiency**

24. The magnesium (Mn) content of a water source averages 0.24 mg/L Mn. What is the percent removal if the treated water averages 0.020 mg/L Mn?

The equation is:

$$\textbf{Percent Mn removal efficiency} = \frac{\textbf{(In – Out)(100\%)}}{\textbf{In}}$$

$$\text{Percent Mn removal efficiency} = \frac{(0.24 - 0.020)(100\%)}{0.24} = \textbf{92\% Mn removal efficiency}$$

PERCENT MIXTURE

25. If 12 lb of soda ash are mixed with 48 gal of water, what is the percent of soda ash in the resulting slurry?

$$\textbf{Percent, soda ash} = \frac{\textbf{(Number of lb, soda ash)(100\%)}}{\textbf{Number of lb, soda ash + (8.34 lb/gal)(Number of gal, water)}}$$

Substituting:

$$\text{Percent soda ash} = \frac{(12 \text{ lb})(100\%)}{12 \text{ lb} + (8.34 \text{ lb/gal})(48 \text{ gal})} =$$

$$\frac{(12 \text{ lb})(100\%)}{12 \text{ lb} + 400.32 \text{ lb}} = \frac{(12 \text{ lb})(100\%)}{412.32 \text{ lb}} = \textbf{2.9\% soda ash}$$

26. What is the percent of lime in the resulting slurry if 45 lb of lime are mixed with exactly 50.0 gal of water?

$$\textbf{Percent lime} = \frac{\textbf{(Number of lb, lime)(100\%)}}{\textbf{Number of lb, lime + (8.34 lb/gal)(Number of gal, water)}}$$

Substituting:

$$\text{Percent lime} = \frac{(45 \text{ lb})(100\%)}{45 \text{ lb} + (8.34 \text{ lb/gal})(50.0 \text{ gal})}$$

$$\text{Percent lime} = \frac{(45 \text{ lb})(100\%)}{45 \text{ lb} + 417 \text{ lb}} = \frac{(45 \text{ lb})(100\%)}{462 \text{ lb}} = \textbf{9.7\% lime}$$

CIRCUMFERENCE, RADII, AND DIAMETER DETERMINATIONS

27. What is the circumference of a tank that is 50.0 ft in diameter?

The equation is:

$$\textbf{Circumference} = \boldsymbol{\pi}\textbf{(Diameter)}$$

Note that $\pi = 3.14$.

$$\text{Circumference} = (3.14)(50.0 \text{ ft}) = \textbf{157 ft}$$

28. A tank has a circumference of 94.2 ft. What is the diameter of the tank?

The equation is:

$$\textbf{Circumference} = \boldsymbol{\pi}\textbf{(Diameter)}$$

Rearrange and solve for the diameter.

$$\textbf{Diameter (D)} = \textbf{(Circumference)}/\boldsymbol{\pi}$$

Substitute the known value.

$$\text{D} = (94.2 \text{ ft})/3.14 = \textbf{30.0 ft}$$

29. If the circumference of a tank is 235.5 ft, what is the radius?

The equation is:

$$\textbf{Circumference} = \mathbf{2\pi r}$$

Rearrange and solve for the radius.

$$r = \text{Circumference}/2\pi$$

Substitute the known value.

$$r = 235.5\ \text{ft}/2(3.14) = \mathbf{37.5\ ft}$$

30. If the radius of a tank is 32 ft, what is the circumference?

The equation is:

$$\textbf{Circumference} = \mathbf{2\pi r}$$

$$\text{Circumference} = 2(3.14)(32) = 200.96\text{, round to } \mathbf{200\ ft}$$

AREA PROBLEMS

31. What is the area of a trench that is 68 ft long and 4.5 ft wide?

The equation is:

$$\textbf{Area} = \textbf{(Length)(Width)} \text{ or } \mathbf{A} = \mathbf{(L)(W)}$$

By substituting length and width, we get:

$$A = (68\ \text{ft})(4.5\ \text{ft}) = 306\ \text{ft}^2\text{, round to } \mathbf{310\ ft^2}$$

32. A sedimentation basin is to be constructed with a length of 375 ft and a width of 88 ft. How much area will the sedimentation basin cover?

The equation is:

$$\mathbf{A = (L)(W)}$$

Substitute known values and solve.

$$A = (375)(88) = \mathbf{33{,}000\ ft^2}$$

33. A trench has an area of 440 ft². If the width is 3.5 ft, what must the length be?

The equation is:

$$\mathbf{A = (L)(W)}$$

Rearrange the equation to solve for length.

$$L = \frac{A}{W}$$

or:

$$L = A/W$$

Substituting:

$$L = 440\ ft^2/3.5\ ft = 125.7\ ft, \text{ round to } \mathbf{130\ ft}$$

34. How much surface area is contained by a tank 40.0 ft in diameter?

The equation is:

$$\mathbf{Area = (0.785)(Diameter)^2}$$

or:

$$\mathbf{Area = \tfrac{1}{2}(r)^2}$$

where r is the radius.

$$\mathbf{r = \tfrac{1}{2}(Diameter)}$$

Using the first equation:

$$\text{Area of tank} = (0.785)(40.0\ ft)(40.0\ ft) = 1{,}256\ ft, \text{ round to } \mathbf{1{,}260\ ft^2}$$

35. What is the cross-sectional area of a pipe that is 14 in. in diameter?

The equation is:

$$\textbf{Area, ft}^2 = \textbf{(0.785)(Diameter)}^2$$

First, convert pipe diameter from inches to feet.

$$14 \text{ in./12 in. per foot} = 1.167$$

$$\text{Area, ft}^2 = (0.785)(1.167 \text{ ft})(1.167 \text{ ft}) = 1.069 \text{ ft}^2\text{, round to } \mathbf{1.1\ ft^2}$$

36. If a circular tank covers an area of 1,962.5 ft, what is the diameter of the tank?

The equation is:

$$\textbf{Area} = \textbf{(0.785)(Diameter)}^2$$

Rearrange and solve for the diameter.

$$(\text{Diameter})^2 = \frac{\text{Area}}{(0.785)} = \frac{1{,}962.5 \text{ ft}^2}{(0.785)} = 2{,}500 \text{ ft}^2$$

Next, take the square root of both sides.

$$\textbf{(Diameter)}^2 = \textbf{50.0 ft}$$

37. What is the exposed exterior surface area of a ground-level storage tank that is 16.25 ft high and has a diameter of 125 ft? Assume the top is flat.

First, find the square footage of the wall area.

$$\textbf{Wall area, ft}^2 = \textbf{(Diameter)}(\pi)\textbf{(Height)}$$

$$\text{Wall area, ft}^2 = (125 \text{ ft})(3.14)(16.25 \text{ ft}) = 6{,}378 \text{ ft}^2$$

Next, find the area of the top. Note: There is no bottom exterior area.

$$\text{Top area, ft}^2 = (0.785)(\text{Diameter})^2 = (0.785)(125 \text{ ft})(125 \text{ ft}) = 12{,}266 \text{ ft}^2$$

$$\text{Total exterior surface area of tank, ft}^2 = 6{,}378 \text{ ft}^2 + 12{,}266 \text{ ft}^2$$

$$\text{Total exterior surface area of tank, ft}^2 = 18{,}644 \text{ ft}^2\text{, round to } \mathbf{18{,}600\ ft^2}$$

38. What is the exterior surface area in square feet of a cylindrical tank that is 18.0 ft high and 112.0 ft in diameter? Assume the top is flat.

First, calculate the top surface area.

The equation is:

$$\textbf{Top of tank, ft}^2 = \textbf{(0.785)(Diameter)}^2$$

$$\text{Top of tank, ft}^2 = (0.785)(112.0 \text{ ft})(112.0 \text{ ft}) = 9{,}847 \text{ ft}^2$$

Next, calculate the tank wall area.

The area of a cylindrical wall is:

$$\text{Wall, ft}^2 = (\text{Diameter})(\pi)(\text{Height of tank})$$

$$\text{Wall, ft}^2 = (112.0 \text{ ft})(3.14)(18.0 \text{ ft}) = 6{,}330 \text{ ft}^2$$

$$\text{Total exterior surface area of tank} = 9{,}847 \text{ ft}^2 + 6{,}330 \text{ ft}^2$$

$$= 16{,}177 \text{ ft}^2\text{, round to } \mathbf{16{,}200 \text{ ft}^2}$$

VOLUME DETERMINATIONS

39. Find the number of gallons in an acre-foot, given the following data:

1 L	=	**0.03532 ft^3**
1 acre-ft	=	**43,560 ft^3**
1 gal	=	**3.785 L**

First, find the number of liters.

$$\text{Number of L} = \text{Number of ft}^3/\text{Number of ft}^3 \text{ per L}$$

$$\text{Number of L} = \frac{43{,}560 \text{ ft}^3}{0.03532 \text{ ft}^3/\text{L}} = 1{,}233{,}295.583 \text{ L}$$

Next, find the number of gallons in an acre-foot.

$$\text{Number of gal} = \frac{1{,}233{,}295.583 \text{ L}}{3.785 \text{ L/gal}} = 325{,}838 \text{ gal, round to } \mathbf{325{,}800 \text{ gal}}$$

40. What is the volume of a trench in cubic feet if the trench is 24 ft by 3.0 ft by 6.0 ft (length, width, and depth, respectively)?

The equation is:

$$\textbf{Volume, ft}^3 = \textbf{(Length)(Width)(Depth)} \text{ or } \textbf{(L)(W)(D)}$$

$$\text{Volume, ft}^3 = (24 \text{ ft})(3.0 \text{ ft})(6.0 \text{ ft}) = 432 \text{ ft}^3\text{, round to } \textbf{430 ft}^3$$

41. What is the volume of a reservoir in gallons if it is 145 ft long, 76 ft wide, and averages 12 ft in depth?

The equation is:

$$\textbf{Volume, gal} = \textbf{(L)(W)(D)(7.48 gal/ft}^3\textbf{)}$$

$$\text{Volume, gal} = (145 \text{ ft})(76 \text{ ft})(12 \text{ ft})(7.48 \text{ gal/ft}^3) = 989{,}155 \text{ gal, round to } \textbf{990,000 gal}$$

42. What is the capacity of a tank in gallons if the diameter is 80.0 ft and the overflow is 32 ft from the base?

The equation is:

$$\textbf{Volume, gal} = \textbf{(0.785)(Diameter)}^2\textbf{(Depth)(7.48 gal/ft}^3\textbf{)}$$

$$\text{Volume, gal} = (0.785)(80.0 \text{ ft})(80.0 \text{ ft})(32 \text{ ft})(7.48 \text{ gal/ft}^3)$$

$$\text{Volume, gal} = 1{,}202{,}545 \text{ gal, round to } \textbf{1,200,000 gal}$$

43. A small tank has a diameter of 2.3 ft and a calcium hypochlorite level of 3.6 ft. How many gallons of calcium hypochlorite are contained in the tank?

The equation is:

$$\textbf{Volume, gal} = \textbf{(0.785)(Diameter)}^2\textbf{(Depth)(7.48 gal/ft}^3\textbf{)}$$

$$\text{Volume, gal} = (0.785)(2.3 \text{ ft})(2.3 \text{ ft})(3.6 \text{ ft})(7.48 \text{ gal/ft}^3) = 111.8 \text{ gal, round to } \textbf{110 gal}$$

44. A trench for a water main has to be 675 ft long, 4 ft wide, and 6 ft deep. How many cubic yards must be excavated?

First, find the number of cubic feet.

The equation is:

$$\textbf{Volume, ft}^3 = \textbf{(L)(W)(D)}$$

$$\text{Volume, ft}^3 = (675 \text{ ft})(4 \text{ ft})(6 \text{ ft}) = 16{,}200 \text{ ft}^3$$

Next, convert cubic feet to cubic yards.

Thus:

$$1 \text{ yd}^3 = 27 \text{ ft}^3$$

$$\frac{16{,}200 \text{ ft}^3}{27 \text{ ft}^3/\text{yd}^3} = \mathbf{600 \text{ yd}^3}$$

45. How many liters of water can be contained in a tank that has a diameter of 2.5 ft and is 6.0 ft high?

The equation is:

$$\textbf{Volume, L} = \textbf{(0.785)(Diameter)}^2\textbf{(Depth)(7.48 gal/ft}^3\textbf{)(3.785 L/gal)}$$

$$\text{Volume, L} = (0.785)(2.5 \text{ ft})(2.5 \text{ ft})(6.0 \text{ ft})(7.48 \text{ gal/ft}^3)(3.785 \text{ L/gal})$$

$$\text{Volume, L} = 833 \text{ L, round to } \mathbf{830 \text{ L}}$$

46. A pipe is 7.26 mi long and has an inner diameter of 24 in. How many gallons can fit in the pipeline?

The equation is:

Volume, gal = (0.785)(Diameter)^2(Length)(7.48 gal/ft^3)

First, convert miles to feet.

(7.26 mi)(5,280 ft/mi) = 38,333 ft

Next, convert 24 in. to feet.

24 in./12 in. per foot = 2.0 ft

Last, substitute known values and solve.

Volume, gal = (0.785)(2.0 ft)(2.0 ft)(38,333 ft)(7.48 gal/ft^3)

Volume, gal = 900,335 gal, round to **900,000 gal**

47. A pipe that is 16 in. in diameter and is 260 ft long can contain how many gallons of water?

The equation is:

Volume, gal = (0.785)(Diameter)^2(Length)(7.48 gal/ft^3)

First, convert 16 in. to feet.

16 in./12 in. per foot = 1.33 ft

Volume, gal = (0.785)(1.33 ft)(1.33 ft)(260 ft)(7.48 gal/ft^3) = **2,700 gal**

48. How many liters are contained in 45 gal of water?

First, we know that there are 3.785 L for every gallon of water

The equation is:

$$\textbf{Number of L} = \textbf{(Number of gallons)(3.785 L/gal)}$$

$$\text{Number of L} = (45 \text{ gal})(3.785 \text{ L/gal}) = \mathbf{170 \text{ L}}$$

49. How many liters are in a pipe that is 10.0 in. in diameter and 365 m long?

First, convert the diameter from inches to meters.

$$\text{Number of m} = \frac{10.0 \text{ in.}}{39.37 \text{ in./m}} = 0.254 \text{ m}$$

Next, calculate the volume.

$$\textbf{Number of L} = \textbf{(0.785)(Diameter)}^2\textbf{(1,000 L/m}^3\textbf{)}$$

$$\text{Number of L} = (0.785)(0.254 \text{ m})(0.254 \text{ m})(365 \text{ m})(1{,}000 \text{ L/m}^3)$$

$$\text{Number of L} = 18{,}485 \text{ L, round to } \mathbf{18{,}500 \text{ L}}$$

50. A small lake used as a water source is found to average 32 ft in depth. The shape of the lake is approximately circular with a diameter of approximately 4,000 ft. How many cubic feet of water does the lake contain?

The equation is:

$$\textbf{Volume, ft}^3 = \textbf{(0.785)(Diameter)}^2\textbf{(Depth)}$$

$$\text{Volume, ft}^3 = (0.785)(4{,}000 \text{ ft})(4{,}000 \text{ ft})(32 \text{ ft})$$

$$= 401{,}920{,}000 \text{ ft}^3\text{, round to } \mathbf{400{,}000{,}000 \text{ ft}^3}$$

51. A trench that averages 3.5 ft wide and 4.0 ft in depth is dug for the purpose of installing a 24-in. diameter pipeline. If the trench is 1,663 ft long, how much soil in cubic feet will be put in the trench after the pipe is in place, assuming that the only soil left over is that which the pipe now occupies?

First, find the volume of soil removed from the trench in cubic feet.

The equation is:

$$\textbf{Volume, ft}^3 = \textbf{(L)(W)(D)}$$

$$\text{Volume, ft}^3 = (1{,}663 \text{ ft})(3.5 \text{ ft})(4.0 \text{ ft}) = 23{,}282 \text{ ft}^3$$

Next, convert 24 in. to feet.

$$24 \text{ in.}/12 \text{ in. per foot} = 2.0 \text{ ft}$$

Next, determine the volume the pipe occupies.

The equation is:

$$\textbf{Volume, ft}^3 = \textbf{(0.785)(Diameter)}^2\textbf{(Length)}$$

$$\text{Volume, ft}^3 = (0.785)(2.0 \text{ ft})(2.0 \text{ ft})(1{,}663 \text{ ft}) = 5{,}222 \text{ ft}^3$$

Last, subtract the volume of the pipe from the volume of the trench.

Volume of soil put in trench = 23,282 ft^3 – 5,222 ft^3 = 18,060 ft^3, round to **18,000 ft^3**

DETENTION TIME CALCULATIONS

52. If a water tank has a volume 1.5 mil gal and the flow from the tank is 4.25 mgd, what is the detention time in hours?

The equation is:

$$\textbf{Detention time, hr} = \frac{\textbf{(Tank volume)(24 hr/d)}}{\textbf{Flow, gpd}}$$

Substitute known values and solve.

First, convert million gallons to gallons.

$$(1.5 \text{ mil gal})(1{,}000{,}000/1 \text{ M}) = 1{,}500{,}000 \text{ gal}$$

And:

$$(4.25 \text{ mgd})(1{,}000{,}000/1 \text{ M}) = 4{,}250{,}000 \text{ gpd}$$

$$\text{Detention time, hr} = \frac{(1{,}500{,}000 \text{ gal})(24 \text{ hr/d})}{4{,}250{,}000 \text{ gpd}} = \mathbf{8.5 \text{ hr}}$$

53. A water tank averages 1,230,000 gal. If the detention time is 4.7 hr, what is the flow in million gallons per day from the tank?

The equation is:

$$\textbf{Flow, gpd} = \frac{\textbf{(Tank volume)(24 hr/d)}}{\textbf{Detention time, hr}}$$

Substituting:

$$\text{Flow, gpd} = \frac{(1{,}230{,}000 \text{ gal})(24 \text{ hr/d})}{4.7 \text{ hr}} = 6{,}280{,}851 \text{ gal}$$

$$\text{Number of mgd} = \frac{6{,}280{,}851 \text{ gal}}{1{,}000{,}000 \text{ gal/1 M}} = 6.28 \text{ mgd, round to } \mathbf{6.3 \text{ mgd}}$$

54. A storage tank has a 50.0 ft radius and averages 23.5 ft in water depth. What is the average detention time in hours for this storage tank if flow through the tank averages 5.77 mgd during a particular month in question?

First, find the diameter.

$$\text{Diameter, ft} = 2(\text{radius}) = 2(50.0 \text{ ft}) = 100 \text{ ft}$$

Then, determine the volume of water in the storage tank.

The equation is:

$$\textbf{Volume, gal} = \mathbf{(0.785)(Diameter)^2(Depth)(7.48 \text{ gal/ft}^3)}$$

$$\text{Average tank volume, gal} = (0.785)(100 \text{ ft})(100 \text{ ft})(23.5 \text{ ft})(7.48 \text{ gal/ft}^3) = 1{,}379{,}873 \text{ gal}$$

Next, convert million gallons per day to gallons per day.

$$\text{Number of gpd} = (5.77 \text{ mgd})(1{,}000{,}000 \text{ gal/1 M}) = 5{,}770{,}000 \text{ gpd}$$

Next, solve for the detention time.

$$\textbf{Detention time, hr} = \frac{\textbf{(Tank volume)(24 hr/d)}}{\textbf{Flow, gpd}}$$

Substituting:

$$\text{Detention time, hr} = \frac{(1{,}379{,}873 \text{ gal})(24 \text{ hr/d})}{5{,}770{,}000 \text{ gpd}} = \mathbf{5.74 \text{ hr}}$$

55. Find the detention time in hours for the following filtration plant that treats 43.5 mgd:

Five flocculation (floc) basins, each 62 ft by 15 ft with a water depth of 12.5 ft
A sedimentation (sed) basin, 474 ft by 68 ft with a water depth of 12.0 ft
Twelve filters, each 45 ft by 34 ft with average water depths of 10.0 ft

First, determine the number of gallons in the floc basins, sed basin, and filters.

The equation is:

Volume, gal = (Length)(Width)(Depth)(7.48 gal/ft³)(Number of basins or filters)

Volume, gal, in floc basins = (62 ft)(15 ft)(12.5 ft)(7.48 gal/ft³)(5 basins) = 434,775 gal
Volume, gal, in sed basin = (474 ft)(68 ft)(12.0 ft)(7.48 gal/ft³)(1 basin) = 2,893,144 gal
Volume, gal, in filters = (45 ft)(34 ft)(10.0 ft)(7.48 gal/ft³)(12 filters) = 1,373,328 gal

4,701,247 gal

Next, convert million gallons per day to gallons per hour.

(43.5 mgd)(1 d/24 hr)(1,000,000/1 M) = 1,812,500 gph

Last, calculate the detention time in hours.

The equation is:

$$\textbf{Detention time} = \frac{\textbf{Volume, gal}}{\textbf{Flow rate, gph}}$$

$$\text{Detention time} = \frac{4{,}701{,}247 \text{ gal}}{1{,}812{,}500 \text{ gph}} = 2.59 \text{ hr, round to } \mathbf{2.6\ hr}$$

WATERSHED SUPPLY CALCULATIONS

56. A watershed is 445 mi^2. Determine the number of million gallons per year and million gallons per day the watershed can supply a utility given that the average rainfall per year is 26.9 in. and 14% of the rainfall is collected.

First, find the number of inches of rainfall collected each year in feet.

$$(26.9 \text{ in.})(14\%/100\%)(1 \text{ ft}/12 \text{ in.}) = 0.314 \text{ ft of water}$$

Then, find the number of gallons in 1 acre that is covered by 0.314 ft of water.

$$(43{,}560 \text{ ft}^2/\text{acre})(7.48 \text{ gal/ft}^3)(0.314 \text{ ft}) = 102{,}310 \text{ gal/acre}$$

Convert this to million gallons per acre.

$$(102{,}310 \text{ gal/acre})(1 \text{ M}/1{,}000{,}000) = 0.1023 \text{ mil gal/acre}$$

Next, determine the number of acres in 445 mi^2.

$$\text{Number of acres} = (640 \text{ acres/mi}^2)(445 \text{ mi}^2) = 284{,}800 \text{ acres}$$

Then, calculate the number of million gallons collected from the watershed.

$$(0.1023 \text{ mil gal/acre})(284{,}800 \text{ acres}) = 29{,}135 \text{ mil gal/yr, round to } \mathbf{29{,}000 \text{ mil gal/yr}}$$

Last, determine the average daily supply the watershed can give.

$$\frac{29{,}135 \text{ mil gal/yr}}{365 \text{ d/yr}} = 79.82 \text{ mgd, round to } \mathbf{80 \text{ mgd}}$$

57. Determine the number of million gallons per year and million gallons per day a watershed can supply given the following data:

The watershed is 716 mi^2.

Average annual rainfall is 21.2 in.

9.0% of the rainfall is collected.

First, find the number of inches of rainfall collected each year in feet.

$$(21.2 \text{ in.})(9.0\%/100\%)(1 \text{ ft}/12 \text{ in.}) = 0.159 \text{ ft of water}$$

Then, find the number of gallons in 1 acre that is covered by 0.159 ft of water.

$$(43{,}560 \text{ ft}^2/\text{acre})(7.48 \text{ gal/ft}^3)(0.159 \text{ ft}) = 51{,}807 \text{ gal/acre}$$

Convert this to million gallons per acre.

$$(51{,}807 \text{ gal/acre})(1 \text{ M}/1{,}000{,}000) = 0.0518 \text{ mil gal/acre}$$

Next, determine the number of acres in 716 mi^2.

$$\text{Number of acres} = (640 \text{ acres/mi}^2)(716 \text{ mi}^2) = 458{,}240 \text{ acres}$$

Then, calculate the number of million gallons collected from the watershed.

$$(0.0518 \text{ mil gal/acre})(458{,}240 \text{ acres}) = 23{,}737 \text{ mil gal/yr, round to } \mathbf{24{,}000 \text{ mil gal/yr}}$$

Last, determine the average daily supply the watershed can give.

$$\frac{23{,}737 \text{ mil gal/yr}}{365 \text{ d/year}} = 65.03 \text{ mgd, round to } \mathbf{65 \text{ mgd}}$$

PRESSURE PROBLEMS

58. What is the pressure head at a fire hydrant in feet if the pressure gauge reads 189 psi?

The equation is:

$$\mathbf{Pressure\ head,\ ft = (Pressure,\ psi)(2.31\ ft/psi)}$$

$$\text{Pressure head, ft} = (189 \text{ psi})(2.31 \text{ ft/psi}) = 436.59 \text{ ft, round to } \mathbf{437 \text{ ft}}$$

59. If the pressure head on a fire hydrant is 210 ft, what is the pressure in psi?

The equation is:

$$\mathbf{Pressure\ head,\ ft = (Pressure,\ psi)(2.31\ ft/psi)}$$

Rearrange to solve for pressure in psi.

$$\mathbf{Pressure,\ psi = \frac{Pressure\ head,\ ft}{2.31\ ft/psi}}$$

$$\text{Pressure, psi} = \frac{210 \text{ ft}}{2.31 \text{ ft/psi}} = \mathbf{91 \text{ psi}}$$

60. What is the pressure at the bottom of a tank with 32 ft of water?

The equation is:

$$\textbf{Pressure, psi} = \frac{\textbf{(Height or depth of water)}}{\textbf{(2.31 ft/psi)}}$$

$$\text{Pressure, psi} = \frac{(32\text{ ft})}{2.31\text{ ft/psi}} = 13.85\text{ psi, round to } \mathbf{14\ psi}$$

61. The pressure at the lowest point of a clear well is 15.4 psi. What is the total water depth of the clear well at its lowest point?

The equation is:

$$\textbf{Pressure, psi} = \frac{\textbf{Depth of water}}{\textbf{2.31 ft/psi}}$$

Rearrange the equation to solve for depth.

$$\text{Depth of water} = (\text{Pressure, psi})(2.31\text{ ft/psi})$$

$$\text{Depth of water} = (15.4\text{ psi})(2.31\text{ ft/psi}) = \mathbf{35.6\ ft}$$

62. The pressure at the bottom of a reservoir is 132 psi. What is the depth at that point?

The equation is:

$$\textbf{Pressure, psi} = \textbf{(2.31 ft/psi)(Depth of water)}$$

$$\text{or } \textbf{Depth} = \textbf{(Pressure, psi)(0.433 psi/ft)}$$

Rearrange the equation and solve for depth.

$$\text{Depth} = \frac{(132\text{ psi})}{(0.433\text{ psi/ft})} = 304.85\text{ ft, round to } \mathbf{305\ ft}$$

63. A rectangular water storage tank is partially underground. The tank is 40.0 ft long, 15.0 ft wide, and 15.2 ft high. What is the upward force of water pressure in pounds lifting up on the tank if the groundwater is 7.54 ft above the base of the tank?

First, calculate the tank's area.

The equation is:

$$\textbf{Area, ft} = \textbf{(L)(W)}$$

$$\text{Area, ft}^2 = (40.0\text{ ft})(15.0\text{ ft}) = 600\text{ ft}^2$$

Next, determine the pressure in pounds per square feet on the tank's bottom.

The equation is:

$$\textbf{Pressure, lb/ft}^2 = \textbf{(Water density, lb/ft}^3\textbf{)(Height, ft)}$$

$$\text{Pressure, lb/ft}^2 = (62.4\text{ lb/ft}^3)(7.54\text{ ft}) = 470.5\text{ lb/ft}^2$$

Last, find the upward force on the bottom of the tank.

The equation is:

$$\textbf{Upward force, lb} = \textbf{(Area, ft}^2\textbf{)(Pressure, lb/ft}^2\textbf{)}$$

Upward force, lb = (600 ft^2)(470.5 lb/ft^2) = 282,300 lb, round to **282,000 lb**

64. A circular water storage tank that is underground has a diameter of 50.0 ft. What is the upward force of water pressure in pounds lifting up on the tank if the groundwater is 6.2 ft above the base of the tank?

First, calculate the tank's area.

The equation is:

$$\textbf{Area, ft} = \textbf{(0.785)(Diameter)}^2$$

$$\text{Area, ft}^2 = (0.785)(50.0\text{ ft})(50.0\text{ ft}) = 1{,}962.5\text{ ft}^2$$

Next, determine the pressure in pounds per square feet on the tank's bottom.

The equation is:

$$\textbf{Pressure, lb/ft}^2 = \textbf{(Water density, lb/ft}^3\textbf{)(Height, ft)}$$

$$\text{Pressure, lb/ft}^2 = (62.4\text{ lb/ft}^3)(6.2\text{ ft}) = 386.88\text{ lb/ft}^2$$

(continued)

Last, find the upward force on the bottom of the tank.

The equation is:

$$\textbf{Upward force, lb} = \textbf{(Area, ft}^2\textbf{)(Pressure, lb/ft}^2\textbf{)}$$

$$\text{Upward force, lb} = (1{,}962.5\ \text{ft}^2)(386.88\ \text{lb/ft}^2) = 759{,}252\ \text{lb, round to } \mathbf{760{,}000\ lb}$$

65. An underground rectangular water storage tank has a length of 38 ft and is 16 ft wide. Determine the upward force of water pressure in pounds lifting up on the tank if the groundwater is 3.36 ft above the base of the tank.

First, calculate the tank's area.

The equation is:

$$\textbf{Area, ft} = \textbf{(L)(W)}$$

$$\text{Area, ft}^2 = (38\ \text{ft})(16\ \text{ft}) = 608\ \text{ft}^2$$

Next, determine the pressure in pounds per square feet on the tank's bottom.

The equation is:

$$\textbf{Pressure, lb/ft}^2 = \textbf{(Water density, lb/ft}^3\textbf{)(Height, ft)}$$

$$\text{Pressure, lb/ft}^2 = (62.4\ \text{lb/ft}^3)(3.36\ \text{ft}) = 209.66\ \text{lb/ft}^2$$

Last, find the upward force on the bottom of the tank.

The equation is:

$$\textbf{Upward force, lb} = \textbf{(Area, ft}^2\textbf{)(Pressure, lb/ft}^2\textbf{)}$$

$$\text{Upward force, lb} = (608\ \text{ft}^2)(209.66\ \text{lb/ft}^2) = 127{,}473\ \text{lb, round to } \mathbf{130{,}000\ lb}$$

66. An underground circular water storage tank has a diameter of 25.5 ft. What is the upward force of water pressure in pounds lifting up on the tank if the groundwater is 2.88 ft above the base of the tank?

First, calculate the tank's area.

The equation is:

$$\textbf{Area, ft} = \textbf{(0.785)(Diameter)}^2$$

$$\text{Area, ft}^2 = (0.785)(25.5\ \text{ft})(25.5\ \text{ft}) = 510.45\ \text{ft}^2$$

Next, determine the pressure in pounds per square feet on the tank's bottom.

The equation is:

$$\textbf{Pressure, lb/ft}^2 = \textbf{(Water density, lb/ft}^3\textbf{)(Height, ft)}$$

$$\text{Pressure, lb/ft}^2 = (62.4 \text{ lb/ft}^3)(2.88 \text{ ft}) = 179.71 \text{ lb/ft}^2$$

Last, find the upward force on the bottom of the tank.

The equation is:

$$\textbf{Upward force, lb} = \textbf{(Area, ft}^2\textbf{)(Pressure, lb/ft}^2\textbf{)}$$

$$\text{Upward force, lb} = (510.45 \text{ ft}^2)(179.71 \text{ lb/ft}^2) = 91{,}733 \text{ lb, round to } \mathbf{91{,}700 \text{ lb}}$$

FLOW PROBLEMS

67. A water tank is being filled by a water line at a rate of 26 gpm. If the tank's capacity is 4,500 gal, how many minutes will it take to fill the tank?

The equation is:

$$\textbf{Time, min} = \textbf{Number of gallons/Gallons per minute}$$

$$\text{Time, min} = 4{,}500 \text{ gal}/26 \text{ gpm} = \mathbf{173 \text{ min}}$$

68. A meter indicates that water is flowing from a fire hydrant at 3.2 ft³/min. How many gallons will flow from the hydrant in 43 min?

First, convert the flow in cubic feet per minute to gallons per minute.

$$\text{Gallons per minute} = (3.2 \text{ ft}^3/\text{min})(7.48 \text{ gal/ft}^3) = 23.936 \text{ gpm}$$

Then, determine the number of gallons that flowed through the fire hydrant.

$$\text{Gallons} = (23.936 \text{ gpm})(43 \text{ min}) = 1{,}029 \text{ gal, round to } \mathbf{1{,}000 \text{ gal}}$$

69. If a water line is flowing at 1.73 ft³/s, what is the flow in gallons per minute?

The equation is:

$$\textbf{Flow, gpm} = \textbf{(Flow, ft}^3\textbf{/s)(7.48 gal/ft}^3\textbf{)(60 s/min)}$$

Substituting:

$$\text{Flow, gpm} = (1.73 \text{ ft}^3\text{/s})(7.48 \text{ gal/ft}^3)(60 \text{ s/min}) = \textbf{776 gpm}$$

70. A water hydrant is flowing at 245 gpm. How many cubic feet per second is this?

The equation is:

$$\textbf{Flow, ft}^3\textbf{/s} = \frac{\textbf{Flow, gpm}}{\textbf{(7.48 gal/ft}^3\textbf{)(60 s/min)}}$$

Substituting:

$$\text{Flow, ft}^3\text{/s} = \frac{245 \text{ gpm}}{(7.48 \text{ gal/ft}^3)(60 \text{ s/min})} = \textbf{0.546 ft}^3\textbf{/s}$$

71. How many gallons per minute are flowing from a water main if the flow rate is 3.5 ft³/s?

The equation is:

$$\textbf{Flow, gpm} = \textbf{(Flow, ft}^3\textbf{/s)(7.48 gal/ft}^3\textbf{)(60 s/min)}$$

Substituting:

$$\text{Flow, gpm} = (3.5 \text{ ft}^3\text{/s})(7.48 \text{ gal/ft}^3)(60 \text{ s/min}) = 1{,}571 \text{ gpm, round to } \textbf{1,600 gpm}$$

72. A water tank with a capacity of 1.20 mil gal is being filled at a rate of 2,140 gpm. How many hours will it take to fill the tank?

First, convert gallons per minute to gallons per hour.

$$(2{,}140 \text{ gpm})(60 \text{ min/hr}) = 128{,}400 \text{ gph}$$

Next, convert 1.20 mil gal to gallons.

$$(1.20 \text{ mil gal})(1{,}000{,}000 \text{ gal/1 M}) = 1{,}200{,}000 \text{ gal}$$

The equation is:

$$\textbf{Time, hr} = \textbf{Number of gallons/gph}$$

$$\text{Time, hr} = 1{,}200{,}000 \text{ gal}/128{,}400 \text{ gph} = \mathbf{9.35\ hr}$$

73. A tank that is 1.2 m in diameter and 2.1 m tall is being filled with water at a rate of 12.4 L/min. How long will it take in minutes to fill the tank?

First, determine the volume of the tank in liters.

The equation is:

$$\textbf{Volume, L} = \mathbf{(0.785)(Diameter)^2(Depth)(1{,}000\ L/m^3)}$$

$$\text{Volume, L} = (0.785)(1.2 \text{ m})(1.2 \text{ m})(2.1 \text{ m})(1{,}000 \text{ L/m}^3) = 2{,}374 \text{ L}$$

Next, divide the flow rate into the number of liters.

$$\text{Time to fill, min} = \frac{2{,}374 \text{ L}}{12.4 \text{ L/min}} = 191 \text{ min, round to } \mathbf{190\ min}$$

DENSITY CALCULATIONS

The density of a substance is the amount of mass per a given volume, usually grams per cubic centimeter.

74. What is the density of a substance that weighs 88.5 g and occupies a space of 75.2 cm^3?

The equation is:

$$\textbf{Density} = \textbf{Mass/Volume}$$

$$\text{Density} = 88.5 \text{ g}/75.2 \text{ cm}^3 = \mathbf{1.18\ g/cm^3}$$

75. A polymer weighs 3.50 lb and occupies 1.398 L. What is the density of the polymer in grams per cubic centimeter?

First, convert the number of pounds to grams.

The following applies:

$$1 \text{ lb} = 454 \text{ g and } 1 \text{ L} = 1{,}000.027 \text{ cm}^3$$

$$\textbf{Number of g} = \textbf{(Number of lb)(454 g/lb)}$$

$$\text{Number of g} = (3.50 \text{ lb})(454 \text{ g/lb}) = 1{,}589 \text{ g}$$

$$\text{Number of cm}^3 = (1{,}000.027 \text{ cm}^3/1 \text{ L})(1.398 \text{ L}) = 1{,}398 \text{ cm}^3$$

$$\textbf{Density} = \textbf{Mass/Volume}$$

$$\text{Polymer density} = 1{,}589 \text{ g}/1{,}398 \text{ cm}^3 = \mathbf{1.14\ g/cm^3}$$

76. An unknown substance has a density of 2.23 g/cm³. How much space will it occupy in cubic centimeters if it weighs 78.5 lb?

First, convert the number of pound to grams.

$$\textbf{Number of g} = \textbf{(Number of lb)(454 g/lb)}$$

$$\text{Number of g} = (78.5 \text{ lb})(454 \text{ g/lb}) = 35{,}639 \text{ g}$$

We know that 2.23 g occupy a space of 1 cm^3. To get the space 35,639 g occupies, we need to divide by its density.

$$\text{Space occupied by the substance} = \frac{35{,}639 \text{ g}}{2.23 \text{ g/cm}^3} = 15{,}982 \text{ cm}^3\text{, round to } \mathbf{16{,}000\ cm^3}$$

SPECIFIC GRAVITY OF LIQUIDS

Specific gravity (sp gr) is a dimensionless number that compares the density of one substance to that of another. Water is the standard for liquids with a specific gravity of 1.0. The density of water can be expressed in several ways: 62.4 lb/ft^3, 8.34 lb/gal, or 1 g/cm^3.

77. What is the specific gravity of an unknown liquid that has a density of 68.4 lb/ft^3?

First, look at the units so that you can compare the density of water to the unknown in similar units. In this case we know that water has a density of 62.4 lb/ft^3. Thus, divide the density of the unknown by the density of water.

The equation is:

Specific gravity = Density of unknown substance/Density of water

$$\text{Specific gravity of unknown substance} = \frac{68.4 \text{ lb/ft}^3}{62.4 \text{ lb/ft}^3} = \mathbf{1.10 \text{ sp gr}}$$

78. Determine the specific gravity of a polymer solution that weighs 10.67 lb/gal.

To find the specific gravity, choose 8.34 lb/gal for comparing water density to the polymer density because they have similar units.

The equation is:

Specific gravity = Density of unknown substance/Density of water

$$\text{Specific gravity for polymer} = \frac{10.67 \text{ lb/gal}}{8.34 \text{ lb/gal}} = \mathbf{1.28 \text{ sp gr}}$$

79. How many pounds per cubic foot will a solution weigh if it has a specific gravity of 1.21?

The equation is:

Specific gravity = Density of unknown substance/Density of water

Rearrange the equation to solve for the density of the liquid in pounds per cubic foot.

$$\text{Density of liquid, lb/ft}^3 = (\text{Specific gravity})(\text{Density of water})$$

$$\text{Density of liquid, lb/ft}^3 = (1.21)(62.4 \text{ lb/ft}^3) = \mathbf{75.5\ lb/ft^3}$$

METER ACCURACY CALCULATIONS

80. A meter being tested reads 2,412 L. A volumetric tank shows that the actual value is 2,445 L. What is the percent accuracy of the meter?

The equation is:

$$\textbf{Meter accuracy, \%} = \frac{\textbf{(Meter reading, gal)(100\%)}}{\textbf{Actual volume, gal}}$$

$$\text{Meter accuracy, \%} = \frac{(2{,}412 \text{ L})(100\%)}{2{,}445 \text{ L}} = \textbf{98.65\% meter accuracy}$$

81. A meter being tested by a laboratory shows a reading of 323.2 ft³. A volumetric tank used to measure the water that flowed through the meter indicates that the actual volume is 2,465 gal. What is the percent accuracy of the meter?

First, convert cubic feet to gallons.

$$\text{Number of gal} = (323.2 \text{ ft}^3)(7.48 \text{ gal/ft}^3) = 2{,}417.536 \text{ gal}$$

The equation is:

$$\textbf{Meter accuracy, \%} = \frac{\textbf{(Meter reading, gal)(100\%)}}{\textbf{Actual volume, gal}}$$

$$\text{Meter accuracy, \%} = \frac{(2{,}417.536 \text{ gal})(100\%)}{2{,}465 \text{ gal}} = \textbf{98.1\% meter accuracy}$$

82. If a meter had an accuracy of 99.1% and the reading on the meter indicated 1,833 L, what is the actual volume?

The equation is:

$$\textbf{Meter accuracy, \%} = \frac{\textbf{(Meter reading, gal)(100\%)}}{\textbf{Actual volume, gal}}$$

Rearrange to solve for the actual volume.

$$\text{Actual volume, L} = \frac{\text{(Meter reading, gal)(100\%)}}{\text{Meter accuracy, \%}}$$

$$\text{Actual volume, L} = \frac{(1{,}833\ \text{L})(100\%)}{99.1\%} = 1{,}849.6\ \text{L, round to } \mathbf{1{,}850\ L}$$

CALCULATIONS FOR CHLORINE DEMAND AND RESIDUAL

83. A treatment plant produces 3.5 mgd with a chlorine dosage of 2.25 mg/L. If a distant storage tank has a chlorine residual of 1.89 mg/L, what has been the chlorine demand of the distribution system and the storage tank?

The equation is:

Chlorine demand, mg/L = Chlorine dose, mg/L – Chlorine residual, mg/L

Chlorine demand, mg/L = 2.25 mg/L – 1.89 mg/L = **0.36 mg/L**

84. A water treatment operator found old records that indicated a chlorine dosage of 2.0 mg/L at the effluent sampling point and a chlorine demand of 0.57 mg/L at a distant point in the distribution system. What must have been the chlorine residual?

Write the equation to solve for the unknown (chlorine residual) by rearranging the equation in problem 83.

Chlorine residual, mg/L = Chlorine dose, mg/L – Chlorine demand, mg/L

Chlorine residual, mg/L = 2.0 mg/L – 0.57 mg/L = **1.43 mg/L**

DOSAGE CALCULATIONS

85. How many pounds per day of chlorine are required to treat 3.40 mgd if the dosage is 1.50 mg/L?

The equation is:

$$\textbf{Pounds per day} = \textbf{(mgd)(Dosage, mg/L)(8.34 lb/d)}$$

$$\text{Chlorine, lb/d} = (3.40 \text{ mgd})(1.50 \text{ mg/L})(8.34 \text{ lb/gal}) = \textbf{42.5 lb/d of chlorine}$$

86. What is the chlorine dosage at a water treatment plant if the chlorinator is set on 155 lb/d and the plant is treating 6.8 mgd?

The equation is:

$$\textbf{Pounds per day} = \textbf{(mgd)(Dosage, mg/L)(8.34 lb/d)}$$

Rearrange the formula and solve for dosage.

$$\text{Dosage, mg/L} = \frac{\text{Pounds per day}}{(\text{mgd})(8.34 \text{ lb/gal})}$$

$$\text{Dosage, mg/L} = \frac{155 \text{ lb/d}}{(6.8 \text{ mgd})(8.34 \text{ lb/gal})} = 2.73 \text{ mg/L, round to } \textbf{2.7 mg/L of chlorine}$$

87. What should the setting be on a chlorinator in pounds per day if the dosage desired is 1.75 mg/L and the pumping rate from the well is 450 gpm?

First, convert the pumping rate to million gallons per day.

The equation is:

$$\textbf{Million gallons per day} = \frac{\textbf{(Pumping rate, gpm)(1,440 min/d)}}{\textbf{1,000,000/M}}$$

Substitute known values and solve.

$$\text{Million gallons per day} = \frac{(450 \text{ gpm})(1{,}440 \text{ min/d})}{1{,}000{,}000/\text{M}} = 0.648 \text{ mgd}$$

Next, use the "pounds" equation to solve the problem.

The equation is:

$$\textbf{Chlorine, lb/d} = \textbf{(mgd)(Dosage, mg/L)(8.34 lb/gal)}$$

$$\text{Chlorine, lb/d} = (0.648 \text{ mgd})(1.75 \text{ mg/L})(8.34 \text{ lb/gal}) = \textbf{9.5 lb/d of chlorine}$$

88. How many gallons of sodium hypochlorite solution that contains 5.0% available chlorine are needed to disinfect a 10.0-in. diameter pipeline that is 220 ft long if the chlorine dosage required is 25 mg/L? Assume the sodium hypochlorite solution has a specific gravity of 1.07.

First, convert 10.0 in. to feet.

$$10.0 \text{ in./12 in. per foot} = 0.833 \text{ ft}$$

Next, find the volume of the pipe.

$$\textbf{Volume, gal} = \textbf{(0.785)(Diameter)}^2\textbf{(Length)(7.48 gal/ft}^3\textbf{)}$$

$$\text{Volume, gal} = (0.785)(0.833 \text{ ft})(0.833 \text{ ft})(220 \text{ ft})(7.48 \text{ gal/ft}^3) = 896 \text{ gal}$$

Next, find the number of million gallons.

$$\text{Million gallons} = (896 \text{ gal})(1 \text{ M}/1{,}000{,}000) = 0.000896 \text{ mil gal}$$

Then, use the "pounds" equation.

$$\textbf{Pounds, sodium hypochlorite solution} = \frac{\textbf{(mil gal)(Dosage, mg/L)(8.34 lb/gal)}}{\textbf{Percent available chlorine/100\%}}$$

$$\text{Pounds, sodium hypochlorite solution} = \frac{(0.000896 \text{ mil gal})(25 \text{ mg/L})(8.34 \text{ lb/gal})}{5.0\%/100\%} = 3.74 \text{ lb}$$

Next, calculate the number of pounds per gallon for the sodium hypochlorite solution.

The equation is:

$$\text{Number of pounds per gallon, NaOCl} = (\text{Specific gravity})(8.34 \text{ lb/gal})$$

$$\text{Number of pounds per gallon, NaOCl} = 1.07(8.34 \text{ lb/gal}) = 8.92 \text{ lb/gal}$$

Last, calculate the number of gallons of sodium hypochlorite.

$$\text{Sodium hypochlorite, gal} = 3.74 \text{ lb}/8.92 \text{ lb/gal} = \textbf{0.42 gal of sodium hypochlorite}$$

89. A 1.2-mil gal storage tank needs to be disinfected with a sodium hypochlorite solution that contains 5.5% available chlorine and weighs 8.95 lb/gal. If the chlorine dosage is to be 40.0 mg/L, how many gallons of sodium hypochlorite are required?

First, determine the number of pounds of chlorine needed using the "pounds" equation.

$$\textbf{Pounds} = \textbf{(mil gal)(Dosage, mg/L)(8.34 lb/gal)}$$

$$\text{Chlorine, lb} = (1.2 \text{ mil gal})(40.0 \text{ mg/L})(8.34 \text{ lb/gal}) = 400.32 \text{ lb}$$

Next, find the number of gallons of sodium hypochlorite.

The equation is:

$$\textbf{Sodium hypochlorite, gal} = \frac{\textbf{(Chlorine, lb)(100\%)}}{\textbf{(8.95 lb/gal)(Percent solution)}}$$

$$\text{Sodium hypochlorite, gal} = \frac{(400.32 \text{ lb of } Cl_2)(100\%)}{(8.95 \text{ lb/gal})(5.5\%)}$$

Sodium hypochlorite, gal = 813.24 gal, round to **810 gal of sodium hypochlorite**

90. How many gallons of a sodium hypochlorite solution that contains 5.8% available chlorine are needed to disinfect a water main that is 36.0 in. in diameter and a pipeline that is 477 ft long if the chlorine dosage required is 30.0 mg/L? Assume the sodium hypochlorite solution weighs 8.97 lb/gal.

First, convert 36.0 in. to feet.

$$36.0 \text{ in./12 in. per foot} = 3.00 \text{ ft}$$

Next, find the volume of the pipe.

$$\textbf{Volume, gal} = \textbf{(0.785)(Diameter)}^2\textbf{(Length)(7.48 gal/ft}^3\textbf{)}$$

$$\text{Volume, gal} = (0.785)(3.00 \text{ ft})(3.00 \text{ ft})(477 \text{ ft})(7.48 \text{ gal/ft}^3) = 25{,}208 \text{ gal}$$

Next, find the number of million gallons.

$$\text{Million gallons} = (25{,}208 \text{ gal})(1 \text{ M}/1{,}000{,}000) = 0.0252 \text{ mil gal}$$

Then, use the "pounds" equation:

$$\textbf{Pounds, sodium hypochlorite} = \frac{\textbf{(mil gal)(Dosage, mg/L)(8.34 lb/gal)}}{\textbf{Percent available chlorine/100\%}}$$

$$\text{Pounds, sodium hypochlorite} = \frac{(0.0252 \text{ mil gal})(30.0 \text{ mg/L})(8.34 \text{ lb/gal})}{5.8\% \text{ available chlorine}/100\%} = 108.7 \text{ lb}$$

Last, calculate the number of gallons of sodium hypochlorite.

$$\text{Sodium hypochlorite, gal} = 108.7 \text{ lb}/8.97 \text{ lb/gal}$$

Sodium hypochlorite, gal = 12.12 gal, round to **12 gal of sodium hypochlorite**

91. A 36-in. pipeline, 885 ft long, was disinfected with calcium hypochlorite tablets with 65.2% available chlorine. What was the chlorine dosage in milligrams per liter if 30.1 lb of calcium hypochlorite was used? Assume that because the hypochlorite is so diluted it weighs 8.34 lb/gal.

First, convert the diameter of the pipeline from inches to feet.

$$\text{Number of ft} = \frac{36 \text{ in.}}{12 \text{ in./ft}} = 3.0 \text{ ft}$$

Next, find the number of gallons by determining the volume of the pipeline.

The equation is:

$$\textbf{Volume of pipe, gal} = \textbf{(0.785)(Diameter)}^2\textbf{(Length)}$$

$$\text{Volume of pipe, gal} = (0.785)(3.0 \text{ ft})(3.0 \text{ ft})(885 \text{ ft})(7.48 \text{ gal/ft}^3) = 46{,}768.89 \text{ gal}$$

Then, convert number of gallons to million gallons.

$$\text{Number of mil gal} = \frac{46{,}768.89 \text{ gal}}{1{,}000{,}000/1 \text{ M}} = 0.0468 \text{ mil gal}$$

Finally, calculate the dosage by rearranging the "pounds" equation.

$$\textbf{Number of lb/d} = \frac{\textbf{(mgd)(Dosage, mg/L)(8.34 lb/gal)(100\%)}}{\textbf{Percent available chlorine}}$$

Rearrange the equation and drop the day on each side of the formula as it is not needed.

$$\text{Dosage, mg/L} = \frac{\text{(Number of pounds)(65.2\% available chlorine)}}{\text{(mil gal)(8.34 lb/gal)(100\% calcium hypochlorite)}}$$

$$\text{Dosage, mg/L} = \frac{\text{(30.1 lb, calcium hypochlorite)(65.2\% available chlorine)}}{\text{(0.0468 mil gal)(8.34 lb/gal)(100\% calcium hypochlorite)}}$$

Dosage, mg/L = 50.28 mg/L, round to **50 mg/L of chlorine**

92. Water from a well is being treated with a sodium hypochlorite solution that contains 5.5% available chlorine and weighs 8.95 lb/gal. The well is pumping water at 775 gpm. What is the chlorine dosage if the chlorinator is pumping at a rate of 226 L/day?

First, convert the production rate of the well pump to million gallons per day.

The equation is:

$$\textbf{Million gallons per day} = \frac{\textbf{(Pumping rate)(1,440 min/d)}}{\textbf{1,000,000/1 M}}$$

$$\text{Million gallons per day} = \frac{(775\text{ gpm})(1{,}440\text{ min/d})}{1{,}000{,}000/1\text{ M}} = 1.12\text{ mgd}$$

Second, convert liters per day to gallons per day.

$$\text{Number of gal} = \frac{226\text{ L/d}}{3.785\text{ L/gal}} = 59.7\text{ gpd}$$

Next, calculate the chlorine usage in pounds per day.

The equation is:

$$\textbf{Chlorine usage, lb/d} = \frac{\textbf{(Hypochlorinator flow, gpd)(\% Cl}_2\textbf{ in hypochlorite solution)(8.95 lb/gal)}}{\textbf{100\%}}$$

$$\text{Chlorine usage, lb/d} = \frac{(59.7\text{ gpd})(5.5\%)(8.95\text{ lb/gal})}{100\%} = 29.39\text{ lb/d}$$

Last, calculate the chlorine dosage using the "pounds" equation.

$$\text{Pounds per day} = (\text{mgd})(\text{Dosage, mg/L})(8.34\text{ lb/gal})$$

Rearrange to solve for dosage.

The equation is:

$$\textbf{Dosage, mg/L} = \frac{\textbf{Pounds per day}}{\textbf{(mgd)(8.34 lb/gal)}}$$

$$\text{Dosage, mg/L} = \frac{29.39\text{ lb/d}}{(1.12\text{ mgd})(8.34\text{ lb/d})} = \textbf{3.1 mg/L, sodium hypochlorite}$$

93. A 2.5-mil gal storage tank needs to be disinfected with a calcium hypochlorite tablet that contains 62.5% available chlorine. The tank is to be filled at 10.0% capacity and the initial chlorine dosage desired is 25.0 mg/L. How many pounds of calcium hypochlorite are required?

First, find the amount of water to be disinfected.

$$10.0\% \text{ capacity} = (2.5 \text{ mil gal})(10.0\%/100\%) = 0.25 \text{ mil gal}$$

Next, determine the number of pounds of chlorine needed using the "pounds" equation.

$$\textbf{Pounds} = \frac{\textbf{(mil gal)(Dosage, mg/L)(8.34 lb/gal)}}{\textbf{Percent available chlorine/100\%}}$$

$$Ca(OCl)_2\text{, lb} = \frac{(0.25 \text{ mil gal})(25.0 \text{ mg/L})(8.34 \text{ lb/gal})}{62.5\%/100\%} = 83.4 \text{ lb, round to } \mathbf{83\ lb\ Ca(OCl)_2}$$

94. What should the chlorinator setting be in kilograms per day if 3.2 mgd of water is treated with a dosage of 1.5 mg/L?

First, convert million gallons per day to gallons per day.

$$\text{Gallons per day} = (3.2 \text{ mgd})(1{,}000{,}000/\text{M}) = 3{,}200{,}000 \text{ gpd}$$

Next, convert million gallons per day to cubic meters per day.

The equation is:

$$\textbf{Flow, m}^3\textbf{/d} = \frac{\textbf{(Flow, gpd)(3.785 L/gal)}}{\textbf{1,000 L/m}^3}$$

$$\text{Flow, m}^3\text{/d} = \frac{(3{,}200{,}000 \text{ gpd})(3.785 \text{ L/gal})}{1{,}000 \text{ L/m}^3} = 12{,}112 \text{ m}^3\text{/d}$$

Then, determine the chlorinator setting in kilograms per day.

The equation is:

$$\textbf{Chlorinator setting, kg/d} = \frac{\textbf{(Flow, m}^3\textbf{/d)(Dose, mg/L)(1,000 L/m}^3\textbf{)}}{\textbf{(1,000 mg/gm)(1,000 gm/kg)}}$$

$$\text{Chlorinator setting, kg/d} = \frac{(12{,}112 \text{ m}^3\text{/d})(1.5 \text{ mg/L})(1{,}000 \text{ L/m}^3)}{(1{,}000 \text{ mg/gm})(1{,}000 \text{ gm/kg})} = \mathbf{18\ kg/d}$$

PUMP DISCHARGE PROBLEMS

95. Determine the number of gallons a pump discharges in 1 hour if it is pumping at a rate of 1,340 gpm.

First, convert number of hours to minutes.

$$(1 \text{ hour})(60 \text{ min/hr}) = 60 \text{ min}$$

Next, find the total discharge in gallons.

The equation is:

Discharge, gal = (Pumping time, min)(Pumping rate, gpm)

Discharge, gal = (60 min)(1,340 gpm) = **80,400 gal**

96. If a pump discharges 840 gpm, how many gallons will it discharge in 4 hr and 20 min?

First, calculate the number of minutes in 4 hr.

$$(4 \text{ hr})(60 \text{ min/hr}) = 240 \text{ min}$$

Next, add the remaining 20 min.

$$240 \text{ min} + 20 \text{ min} = 260 \text{ min}$$

Total discharge, gal = (Pumping time, min)(Pumping rate, gpm)

Total discharge, gal = (260 min)(840 gpm) = 218,400 gal, round to **220,000 gal**

PUMPING CALCULATIONS

97. What is the total head in feet for a pump with a total static head of 16 ft and a head loss of 2.5 ft?

The equation is:

Total head, ft = Total static head, ft + Head losses, ft

Total head = 16 ft + 2.5 ft = **18.5 ft**

98. What must be the total static head for a pump with a total head of 26 ft and head losses of 6.5 ft?

The equation is:

Total head, ft = Total static head, ft + Head losses, ft

Rearrange and solve for total static head.

Total static head, ft = Total head, ft – Head losses, ft

Total static head, ft = 26 ft – 6.5 ft = **19.5 ft**

99. Water is being pumped from a water plant that has an elevation of 1,945 ft to an elevation of 2,167 ft. If friction and minor head losses are 24.5 ft, what is the total head?

The equation is:

Total head, ft = Total static head, ft (difference in elevation) + Head losses, ft

Total head, ft = (2,167 ft – 1,945 ft) + 24.5 ft = 222 ft + 24.5 ft

= 246.5 ft, round to **246 ft**

100. What is the motor horsepower (mhp) for a pump with the following parameters?

Motor efficiency:	**91%**	**Total head:**	**98 ft**
Pump efficiency:	**81%**	**Flow:**	**2.44 mgd**

First, convert million gallons per day to gallons per minute.

Gallons per minute = (2.44 mgd)(1,000,000/1 M)(1 day/1,440 min) = 1,694 gpm

The equation is:

$$\textbf{Motor horsepower} = \frac{\textbf{(Flow, gpm)(Total head, ft)}}{\textbf{(3,960)(Motor efficiency)(Pump efficiency)}}$$

The 3,960 is a constant.

Change percent motor and pump efficiencies to decimal form by dividing by 100%.

Motor efficiency: 91%/100% = 0.91

Pump efficiency: 81%/100% = 0.81

$$\text{Motor horsepower} = \frac{(1{,}694 \text{ gpm})(98 \text{ ft})}{(3{,}960)(0.91)(0.81)} = 56.87 \text{ mhp, round to } \mathbf{57\ mph}$$

101. Determine the motor horsepower for a pump with the following parameters:

Motor efficiency:	**90.5%**	**Horsepower (hp):**	**38**
Pump efficiency:	**79.5%**		

Note: The 38 hp in this problem is called the water horsepower (whp). The whp is defined as the actual energy (horsepower) available to pump water. The equation for determining motor horsepower in this problem is different than the equation in problem 100.

The equation to use for the data given in this problem is:

$$\textbf{Motor horsepower} = \frac{\textbf{(Water horsepower)}}{\textbf{(Percent motor efficiency/100\%)(Percent pump efficiency/100\%)}}$$

$$\text{Motor horsepower} = \frac{38 \text{ whp}}{(90.5\%/100\%)(79.5\%/100\%)} = \mathbf{53\ mhp}$$

102. What is the motor horsepower for a pump with the following parameters?

Motor efficiency:	**89%**	**Total head:**	**144 ft**
Pump efficiency:	**78%**	**Flow:**	**1.88 mgd**

First, convert million gallons per day to gallons per minute.

Gallons per minute = (1.88 mgd)(1,000,000/1 M)(1 day/1,440 min) = 1,306 gpm

The equation is:

$$\textbf{Motor horsepower} = \frac{\textbf{(Flow, gpm)(Total head, ft)}}{\textbf{(3,960)(Motor efficiency)(Pump efficiency)}}$$

Change percent motor and pump efficiencies to decimal form by dividing by 100%.

Motor efficiency: **89%/100% = 0.89**

Pump efficiency: **78%/100% = 0.78**

$$\text{Motor horsepower} = \frac{(1{,}306 \text{ gpm})(144 \text{ ft})}{(3{,}960)(0.89)(0.78)} = 68.4 \text{ mhp, round to } \mathbf{68\ mhp}$$

103. Determine the brake horsepower (bhp) if the motor has an efficiency of 88% and the horsepower is 45.

The equation is:

Brake horsepower = (Horsepower)(Percent motor efficiency)

Change the motor efficiency to decimal form: 88%/100% = 0.88

Brake horsepower = (45 hp)(0.88) = 39.6 bhp, round to **40 bhp**

SPECIFIC YIELD

104. A well that produces an average of 235 gpm has a drawdown of 18.5 ft. Find the specific yield in gallons per minute per foot.

The equation is:

$$\textbf{Specific yield, gpm/ft} = \frac{\textbf{Well yield, gpm}}{\textbf{Drawdown, ft}}$$

$$\text{Specific yield, gpm/ft} = \frac{235 \text{ gpm}}{18.5 \text{ ft}} = \textbf{12.7 gpm/ft}$$

105. What is the drawdown of a well if the well yields 198 gpm and the specific yield is 10.2?

Write the equation, arranging it to solve for the unknown, drawdown.

The equation is:

$$\textbf{Drawdown, ft} = \frac{\textbf{Well yield, gpm}}{\textbf{Specific yield, gpm/ft}}$$

$$\text{Drawdown, ft} = \frac{198 \text{ gpm}}{10.2 \text{ gpm/ft}} = \textbf{19.4 ft}$$

WATER CONSUMPTION

106. A water plant serves 59,400 people. If it treats a yearly average of 7.82 mgd, what are the gallons per capita per day?

The following applies:

$$\text{A capita} = 1 \text{ person}$$

First, convert 7.82 mgd to million gallons.

$$\text{Number of gal} = (7.82 \text{ mgd})(1{,}000{,}000 \text{ gal/M}) = 7{,}820{,}000 \text{ gal}$$

$$\textbf{Gallons per capita per day (GPCPD)} = \frac{\textbf{Volume, gal/day}}{\textbf{Population served/day}}$$

$$\text{GPCPD} = \frac{7{,}820{,}000 \text{ gal/day}}{59{,}400 \text{ capita/day}} = \textbf{132 gal/capita} \text{ or } \textbf{132 gallons per person}$$

Distribution Grade 2 Test

(answers on p. 99)

1. What is the internal surface area of a cylindrical tank (bottom, top, and the cylinder wall) if it is 80.0 ft in diameter and 24.0 ft high?

2. A trench that averages 4.6 ft wide and 5.7 ft in depth is dug for the purpose of installing a 24-in. diameter pipeline. If the trench is 805 ft long, how much soil in cubic feet will be put in the trench after the pipe is in place, assuming that the only soil left over is that which the pipe now occupies?

3. A rectangular water storage tank is partially underground. The tank is 55.0 ft long, 28.0 ft wide, and 18.5 ft high. What is the upward force of water pressure in pounds lifting up on the tank if the groundwater is 8.30 ft above the base of the tank?

4. A tank 5.50 m in diameter and 4.25 m tall is being filled with water at a rate of 63.8 L per minute. How long will it take in hours to fill the tank?

5. An unknown substance has a density of 2.86 g/cm^3. How much space will it occupy in cubic centimeters if it weighs 22.8 lb?

6. How many gallons of a sodium hypochlorite solution that contains 6.75% available chlorine are needed to disinfect a 14.0-in. diameter pipeline that is 925 ft long if the dosage required is 50.0 mg/L? Assume the sodium hypochlorite solution weighs 9.12 lb/gal.

7. A 24-in. pipeline 1,440 ft long was disinfected with calcium hypochlorite tablets with 60.5% available chlorine. What was the chlorine dosage in milligrams per liter if 23.4 lb of calcium hypochlorite was used?

8. A storage tank that has a diameter of 120 ft and a water height of 35.5 ft needs to be disinfected with a 60.25% calcium hypochlorite solution that contains 60.25% available chlorine. If the chlorine dosage is to be 25 mg/L, how many gallons of calcium hypochlorite are required?

9. A storage tank has a radius of 60.0 ft and averages 27.4 ft in water depth. What is the average detention time in hours for this storage tank if flow through the tank averages 3.14 mgd during a particular month in question?

10. What should the chlorinator setting be in kilograms per day if 7.45 mgd of water is treated with a dosage of 2.0 mg/L?

Solutions to Distribution Grade 2 Test

1. First, find the surface area of the tank's wall. Imagine cutting the wall and rolling it out so that it is flat. The length of this now flat wall is simply the circumference of the tank, and is defined as:

$$(\text{Diameter})(\pi)$$

The area of the tank wall then is the length of the wall times the height.

$$\textbf{Wall area, ft}^2 = \textbf{(Diameter)}(\pi)\textbf{(Height)}$$

Substituting:

$$\text{Wall area, ft}^2 = (80.0 \text{ ft})(3.14)(24.0 \text{ ft}) = 6{,}028.8 \text{ ft}^2$$

Next, find the top and bottom surface area.

$$\textbf{Top and bottom area, ft}^2 = \textbf{(2, top and bottom)(0.785)(Diameter)}^2$$

$$\text{Top and bottom area, ft}^2 = (2)(0.785)(80.0 \text{ ft})(80.0 \text{ ft}) = 10{,}048 \text{ ft}^2$$

$$\text{Total area of tank, ft}^2 = 6{,}028.8 \text{ ft}^2 + 10{,}048 \text{ ft}^2 = 16{,}076.8 \text{ ft}^2, \text{ round to } \textbf{16,100 ft}^2$$

2. First, find the volume of soil removed from the trench in cubic feet.

The equation is:

$$\textbf{Volume, ft}^3 = \textbf{(Length)(Width)(Depth)}$$

$$\text{Volume, ft}^3 = (805 \text{ ft})(4.6 \text{ ft})(5.7 \text{ ft}) = 21{,}107 \text{ ft}^3$$

Next, convert 24 in. to feet.

$$24 \text{ in./12 in. per foot} = 2.0 \text{ ft}$$

(continued)

Next, determine the volume that the pipe occupies.

The equation is:

$$\textbf{Volume, ft}^3 = \textbf{(0.785)(Diameter)}^2\textbf{(Length)}$$

$$\text{Volume, ft}^3 = (0.785)(2.0 \text{ ft})(2.0 \text{ ft})(805 \text{ ft}) = 2{,}528 \text{ ft}^3$$

Last, subtract the volume of the pipe from the volume of the trench.

$$\text{Volume of soil put in trench} = 21{,}107 \text{ ft}^3 - 2{,}528 \text{ ft}^3 = 18{,}579 \text{ ft}^3\text{, round to } \mathbf{19{,}000 \text{ ft}^3}$$

3. First, calculate the tank's area.

The equation is:

$$\textbf{Area, ft} = \textbf{(Length)(Width)}$$

$$\text{Area, ft} = (55.0 \text{ ft})(28.0 \text{ ft}) = 1{,}540 \text{ ft}^2$$

Next, determine the pressure in pounds per square foot on the tank's bottom.

The equation is:

$$\textbf{Pressure, lb/ft}^2 = \textbf{(Water density, lb/ft}^3\textbf{)(Height, ft)}$$

$$\text{Pressure, lb/ft}^2 = (62.4 \text{ lb/ft}^3)(8.30 \text{ ft}) = 517.92 \text{ lb/ft}^2$$

Last, find the upward force on the bottom of the tank.

The equation is:

$$\textbf{Upward force, lb} = \textbf{(Area, ft}^2\textbf{)(Pressure, lb/ft}^2\textbf{)}$$

$$\text{Upward force, lb} = (1{,}540 \text{ ft}^2)(517.92 \text{ lb/ft}^2) = 797{,}597 \text{ lb, round to } \mathbf{798{,}000 \text{ lb}}$$

4. First, determine the volume in liters of the tank.

The equation is:

$$\textbf{Volume, L} = \textbf{(0.785)(Diameter)}^2\textbf{(Depth)(1,000 L/m}^3\textbf{)}$$

$$\text{Volume, L} = (0.785)(5.50 \text{ m})(5.50 \text{ m})(4.25 \text{ m})(1{,}000 \text{ L/m}^3) = 100{,}922 \text{ L}$$

Next, divide the flow rate into the number of liters.

$$\text{Time to fill, hr} = \frac{(100{,}922 \text{ L})(1 \text{ hr})}{(63.8 \text{ L/min})(60 \text{ min})} = 26.36 \text{ hr, round to } \mathbf{26.4 \text{ hr}}$$

5. First, convert the number of pound to grams.

$$\text{Number of g} = (\text{Number of lb})(454 \text{ g/lb})$$

$$\text{Number of g} = (22.8 \text{ lb})(454 \text{ g/lb}) = 10{,}351 \text{ g}$$

We know that 2.86 g occupy a space of 1 cm^3. To get the space 10,351 g occupies, we need to divide by its density.

$$\text{Space occupied by the substance} = \frac{10{,}351 \text{ g}}{2.86 \text{ g/cm}^3} = 3{,}619 \text{ cm}^3\text{, round to } \mathbf{3{,}620 \text{ cm}^3}$$

6. First, convert 14.0 in. to feet.

$$14.0 \text{ in./12 in. per foot} = 1.167 \text{ ft}$$

Next, find the volume of the pipe.

$$\mathbf{Volume, gal = (0.785)(Diameter)^2(Length)(7.48 \text{ gal/ft}^3)}$$

$$\text{Volume, gal} = (0.785)(1.167 \text{ ft})(1.167 \text{ ft})(925 \text{ ft})(7.48 \text{ gal/ft}^3) = 7{,}397 \text{ gal}$$

Next, find the number of million gallons.

$$\text{Million gallons} = (7{,}397 \text{ gal})(1 \text{ M}/1{,}000{,}000) = 0.0074 \text{ mil gal}$$

Then, use the "pounds" equation:

$$\mathbf{Pounds, sodium hypochlorite solution = \frac{(mil gal)(Dosage, mg/L)(8.34 lb/gal)}{Percent available chlorine/100\%}}$$

$$\text{Pounds, sodium hypochlorite solution} = \frac{(0.0074 \text{ mil gal})(50.0 \text{ mg/L})(8.34 \text{ lb/gal})}{6.75\% \text{ available chlorine}/100\%} = 45.7 \text{ lb}$$

Last, calculate the number of gallons of sodium hypochlorite.

Sodium hypochlorite, gal = 45.7 lb/9.12 lb/gal = **5.0 gal of sodium hypochlorite**

7. First, convert the diameter of the pipeline from inches to feet.

$$\text{Number of feet} = \frac{24 \text{ in.}}{12 \text{ in./ft}} = 2.0 \text{ ft}$$

Next, find the number of gallons by determining the volume of the pipeline.

(continued)

The equation is:

$$\textbf{Volume of pipe, gal} = \mathbf{(0.785)(Diameter)^2(Length)}$$

$$\text{Volume of pipe, gal} = (0.785)(2.0 \text{ ft})(2.0 \text{ ft})(1{,}440 \text{ ft})(7.48 \text{ gal/ft}^3) = 33{,}822 \text{ gal}$$

Then, convert number of gallons to million gallons.

$$\text{Number of mil gal} = \frac{33{,}822 \text{ gal}}{1{,}000{,}000/1 \text{ M}} = 0.0338 \text{ mil gal}$$

Finally, calculate the dosage by rearranging the "pounds" equation.

$$\textbf{Number of lb/d} = \frac{\mathbf{(mgd)(Dosage, mg/L)(8.34 lb/gal)(100\%)}}{\textbf{Percent available chlorine}}$$

Rearrange the formula and drop the day on each side of the formula as it is not needed.

$$\text{Dosage, mg/L} = \frac{(\text{Number of pounds})(60.5\% \text{ available chlorine})}{(\text{mil gal})(8.34 \text{ lb/gal})(100\%)}$$

$$\text{Dosage, mg/L} = \frac{(23.4 \text{ lb, calcium hypochlorite})(60.5\% \text{ available chlorine})}{(0.0338 \text{ mil gal})(8.34 \text{ lb/gal})(100\%)}$$

Dosage, mg/L = 50.2 mg/L, round to **50 mg/L**

8. First, determine the volume in gallons of water in the storage tank.

The equation is:

$$\textbf{Volume, gal} = \mathbf{(0.785)(Diameter)^2(Height)(7.48 gal/ft^3)}$$

$$\text{Volume, gal} = (0.785)(120 \text{ ft})(120 \text{ ft})(35.5 \text{ ft})(7.48 \text{ gal/ft}^3) = 3{,}001{,}664 \text{ gal}$$

Then, convert gallons to million gallons.

$$\text{Number of mil gal} = (3{,}001{,}664 \text{ gal})(1 \text{ M}/1{,}000{,}000) = 3.0 \text{ mil gal}$$

Next, determine the number of pounds of chlorine needed using the "pounds" equation.

$$\textbf{Pounds} = \mathbf{(mil gal)(Dosage, mg/L)(8.34 lb/gal)}$$

$$\text{Calcium hypochlorite, lb} = (3.0 \text{ mil gal})(25 \text{ mg/L})(8.34 \text{ lb/gal}) = 625.5 \text{ lb}$$

Next, find the number of gallons of calcium hypochlorite.

The equation is:

$$\textbf{Calcium hypochlorite, gal} = \frac{\textbf{(Chlorine, lb)(100\%)}}{\textbf{(8.34 lb/gal)(Percent, solution)}}$$

$$\text{Calcium hypochlorite, gal} = \frac{(625.5 \text{ lb } Cl_2)(100\%)}{(8.34 \text{ lb/gal})(60.25\%)} = 124.48 \text{ gal, round to } \mathbf{124\ gal}$$

9. First, find the diameter.

$$\text{Diameter, ft} = 2(\text{radius}) = 2(60.0 \text{ ft}) = 120 \text{ ft}$$

Then, determine the volume of water in the storage tank.

The equation is:

$$\textbf{Volume, gal} = \textbf{(0.785)(Diameter)}^2\textbf{(Depth)(7.48 gal/ft}^3\textbf{)}$$

$$\text{Average tank volume, gal} = (0.785)(120 \text{ ft})(120 \text{ ft})(27.4 \text{ ft})(7.48 \text{ gal/ft}^3) = 2{,}316{,}777 \text{ gal}$$

Next, convert million gallons per day to gallons per day.

$$\text{Number of gpd} = (3.14 \text{ mgd})(1{,}000{,}000 \text{ gal/M}) = 3{,}140{,}000 \text{ gpd}$$

Next, solve for the detention time.

$$\textbf{Detention time, hr} = \frac{\textbf{(Tank volume)(24 hr/d)}}{\textbf{Flow, gpd}}$$

Substituting:

$$\text{Detention time, hr} = \frac{(2{,}316{,}777 \text{ gal})(24 \text{ hr/d})}{3{,}140{,}000 \text{ gpd}} = \mathbf{17.7\ hr}$$

10. First, convert million gallons per day to gallons per day.

$$\text{Gallons per day} = (7.45 \text{ mgd})(1{,}000{,}000/\text{M}) = 7{,}450{,}000 \text{ gpd}$$

Next, convert million gallons per day to cubic meters per day.

The equation is:

$$\textbf{Flow, m}^3\textbf{/d} = \frac{\textbf{(Flow, gpd)(3.785 L/gal)}}{\textbf{1,000 L/m}^3}$$

$$\text{Flow, m}^3\text{/d} = \frac{(7{,}450{,}000 \text{ gpd})(3.785 \text{ L/gal})}{1{,}000 \text{ L/m}^3} = 28{,}198 \text{ m}^3\text{/d}$$

(continued)

Then, determine the chlorinator setting in kilograms per day.

The equation is:

$$\textbf{Chlorinator setting, kg/d} = \frac{\textbf{(Flow, m}^3\textbf{/d)(Dose, mg/L)(1,000 L/m}^3\textbf{)}}{\textbf{(1,000 mg/g)(1,000 g/kg)}}$$

$$\text{Chlorinator setting, kg/d} = \frac{(28{,}198 \text{ m}^3\text{/d})(2.0 \text{ mg/L})(1{,}000 \text{ L/m}^3)}{(1{,}000 \text{ mg/g})(1{,}000 \text{ g/kg})} = \textbf{56 kg/d, Cl}_2$$

CHAPTER 3

DISTRIBUTION

Grade 3

Students preparing for the grade 4 distribution certification test should also understand these problems.

CONVERSION PROBLEMS

1. Convert 1,000 gal of water to pounds.

Water weighs 8.34 lb/gal, thus

$$\text{Number of lb, water} = (\text{Number of gal})(8.34 \text{ lb/gal})$$

$$\text{Number of lb, water} = (1{,}000 \text{ gal})(8.34 \text{ lb/gal}) = \mathbf{8{,}340 \text{ lb}}$$

2. Convert 152 L of water to pounds.

$$\text{Number of lb} = (\text{Number of L})(1 \text{ gal}/3.785 \text{ L})(8.34 \text{ lb/gal})$$

$$\text{Number of lb} = (152 \text{ L})(1 \text{ gal}/3.785 \text{ L})(8.34 \text{ lb/gal}) = \mathbf{335 \text{ lb}}$$

3. Convert 1,288 ft to meters.

The following applies:

$$1 \text{ m} = 3.281 \text{ ft}$$

$$\text{Number of m} = \frac{1{,}288 \text{ ft}}{3.281 \text{ ft/m}} = \mathbf{392.6 \text{ m}}$$

4. **Convert 75 L/s to gallons per minute.**

The equation is:

$$\textbf{Number of gpm} = \textbf{(Number of L/s)(60 s/min)(1 gal/3.785 L)}$$

$$\text{Number of gpm} = (75 \text{ L/s})(60 \text{ s/min})(1 \text{ gal}/3.785 \text{ L})$$

$$= 1{,}188.9 \text{ gpm, round to } \mathbf{1{,}200 \text{ gpm}}$$

4. **Convert 2,800 gpm to cubic meters per second.**

First, convert gallons to liters.

The equation is:

$$\textbf{Number of L} = \textbf{(Number of gpm)(1 min/60 s)(3.785 L/gal)}$$

$$\text{Number of L} = (2{,}800 \text{ gpm})(1 \text{ min}/60 \text{ s})(3.785 \text{ L/gal}) = 176.6 \text{ L/s}$$

Next, convert liters per second to cubic meters per second.

The equation is:

$$\text{Number of m}^3\text{/s} = \frac{\text{Flow, L/s}}{1{,}000 \text{ L/m}^3}$$

$$\text{Number of m}^3\text{/s} = \frac{176.6 \text{ L/s}}{1{,}000 \text{ L/m}^3} = 0.1766 \text{ m}^3\text{/s, round to } \mathbf{0.18 \text{ m}^3\text{/s}}$$

TEMPERATURE CONVERSIONS

6. **Convert 43°C to degrees Fahrenheit.**

The equation for Fahrenheit is:

$$\mathbf{°F = (9/5)°C + 32}$$

$$°F = (9/5)43°C + 32 = 109.4°F\text{, round to } \mathbf{110°F}$$

7. Convert 170°F to degrees Celsius.

The equation for Celsius is:

$$°C = \frac{5}{9}(°F - 32)$$

$$°C = \frac{5}{9}(170°F - 32) = \frac{5}{9}(138°F) = 76.67°F\text{, round to } \mathbf{77°F}$$

PERCENT AND PARTS PER MILLION CONVERSION PROBLEMS

8. Convert a solution that has 68,000 ppm to percent.

The equation is:

$$\frac{\textbf{(Number of ppm)(100\%)}}{\textbf{1,000,000/1 M}}$$

$$\text{Percent} = \frac{(68{,}000 \text{ ppm})(100\%)}{1{,}000{,}000/1 \text{ M}} = \mathbf{6.8\% \text{ solution}}$$

9. Convert 3,740 ppm to percent.

$$\text{Percent} = \frac{(3{,}740 \text{ ppm})(100\%)}{1{,}000{,}000/1 \text{ M}} = \mathbf{0.374\%}$$

10. What is the percent polymer in a tank with a concentration of 17.3 mg/L?

The following applies:

$$1 \text{ mg/L} = 1 \text{ ppm}$$

Thus, 17.3 mg/L = 17.3 ppm.

The equation is:

$$\textbf{Percent solution} = \frac{\textbf{Number of ppm}}{\textbf{10,000 ppm/1\%}}$$

$$\text{Percent polymer} = \frac{17.3 \text{ ppm}}{10{,}000 \text{ ppm/1\%}} = \mathbf{0.00173\%}$$

PERCENT CALCULATIONS

11. If 45 is 79%, what is 100%?

First, change 79% to decimal form by dividing.

$$79\%/100\% = 0.79$$

Next, let x = the unknown number.

$$x = 45/0.79 = \mathbf{57}$$

12. What is 85% of 478?

First, change percent to decimal form by dividing by 100%.

$$(85\%/100\%) = 0.85$$

Next, let x = the unknown number.

$$x = (478)(0.85) = 406.3\text{, round to } \mathbf{406}$$

13. If 18.5 lb of soda ash are mixed into 53 gal of water, what is the percent of soda ash in the slurry?

First, find the number of pounds in 53 gal of water.

$$\text{Number of lb} = (8.34\text{ lb/gal})(53\text{ gal}) = 442\text{ lb}$$

Next, determine the percent soda ash.

$$\text{Percent soda ash} = \frac{(18.5\text{ lb})(100\%)}{18.5\text{ lb} + 442\text{ lb}} = \frac{(18.5\text{ lb})(100\%)}{460.5\text{ lb}} = \mathbf{4.02\%\ soda\ ash}$$

14. What is the percent lime in a slurry if 28 lb of lime are mixed in a 55-gal drum that contains 48 gal of water?

$$\text{Percent lime} = \frac{(28\text{ lb})(100\%)}{28 + (8.34\text{ lb/gal})(48\text{ gal})} = \frac{(28\text{ lb})(100\%)}{(28\text{ lb} + 400.32\text{ lb})} =$$

$$\frac{(28\text{ lb})(100\%)}{428.32\text{ lb}} = \mathbf{6.5\%\ lime}$$

PERCENT REMOVAL CALCULATIONS

15. What is the iron removal efficiency through a water plant if the source water iron content is 4.25 mg/L and the treated water entering the distribution system is 0.030 mg/L?

The equation is:

$$\textbf{Percent iron removal efficiency} = \frac{\textbf{(Influent, mg/L – Effluent, mg/L)(100\%)}}{\textbf{Influent mg/L}}$$

Condensed version of above equation:

$$\text{Percent iron removal efficiency} = \frac{(\text{In} - \text{Out})(100\%)}{\text{In}}$$

Percent ntu removal efficiency =

$$\frac{(4.25\text{ mg/L} - 0.030\text{ mg/L})(100\%)}{4.25\text{ mg/L}} = \mathbf{99.3\%\ iron\ removal\ efficiency}$$

16. What is the turbidity removal efficiency through a water plant if the source water turbidity is 18.8 ntu and the treated water entering the distribution system is 0.035 ntu?

The equation is:

$$\textbf{Percent ntu removal efficiency} = \frac{\textbf{(In – Out)(100\%)}}{\textbf{In}}$$

Percent ntu removal efficiency =

$$\frac{(18.8\text{ ntu} - 0.035\text{ ntu})(100\%)}{18.8\text{ ntu}} = \mathbf{99.8\%\ ntu\ removal\ efficiency}$$

17. If the influent turbidity for a water plant is 17.5 ntu and the effluent turbidity is 0.03, what is the percent removal?

The equation is:

$$\textbf{Percent ntu removal} = \frac{\textbf{(Influent ntu – Effluent ntu)}}{\textbf{Influent ntu}} \times \textbf{100\%}$$

Or, more simply:

$$\text{Percent removal} = \frac{(\text{In} - \text{Out})}{\text{In}} \times 100\%$$

$$\text{Percent ntu removal} = \frac{(17.5 \text{ ntu} - 0.03 \text{ ntu})(100\%)}{17.5 \text{ ntu}} = \textbf{99.8\% ntu removal}$$

18. What is the percent removal across a settling basin if the influent turbidity is 8.8 ntu and effluent turbidity at the settling basin is 0.89 ntu?

The equation is:

$$\textbf{Percent ntu removal} = \frac{\textbf{(In – Out)}}{\textbf{In}} \times \textbf{100\%}$$

$$\text{Percent ntu removal} = \frac{(8.8 \text{ ntu} - 0.89 \text{ ntu})(100\%)}{8.8 \text{ ntu}} = \textbf{90\% ntu removal}$$

AREA PROBLEMS

19. What is the area of a circular tank that has a radius of 22.0 ft?

First, find the diameter of the tank.

The equation is:

$$\textbf{Diameter, ft} = \textbf{2(Radius)}$$

$$\text{Diameter, ft} = 2(22.0 \text{ ft}) = 44.0 \text{ ft}$$

Next, find the area of the tank.

The equation is:

$$\textbf{Area, ft}^2 = \textbf{(0.785)(Diameter)}^2$$

$$\text{Area, ft}^2 = (0.785)(44.0 \text{ ft})(44.0 \text{ ft}) = \mathbf{1{,}520\ ft^2}$$

20. The circumference of a tank is 188.5 ft. What is the tank's diameter and area?

First, solve for the diameter.

The equation is:

$$\textbf{Circumference} = (\pi)\textbf{(Diameter)}$$

Rearrange the equation to solve.

$$\text{Diameter} = \frac{\text{Circumference}}{(\pi)}$$

$$\text{Diameter} = \frac{188.5 \text{ ft}}{3.14} = \mathbf{60.0\ ft}$$

Last, determine the area of the tank.

$$\textbf{Area, ft}^2 = \textbf{(0.785)(Diameter)}^2$$

$$\text{Area, ft}^2 = (0.785)(60.0 \text{ ft})(60.0 \text{ ft}) = 2{,}826 \text{ ft}^2\text{, round to } \mathbf{2{,}830\ ft^2}$$

21. Determine the area of four flocculation basins and a sedimentation basin given the following:

Flocculation basins: 25 ft by 45 ft each

Sedimentation basin: 65 ft by 310 ft

First, find the area of the flocculation basins.

The equation is:

$$\textbf{Area, ft}^2 = \textbf{(W)(L)(Number of basins)}$$

$$\text{Area, ft}^2 = (25 \text{ ft})(45 \text{ ft})(4 \text{ basins}) = 4{,}500 \text{ ft}^2$$

Next, calculate the area for the sedimentation basin.

$$\text{Area, ft}^2 = (65 \text{ ft})(310 \text{ ft}) = 20{,}150 \text{ ft}^2$$

Now, add the two areas to get the total area that the four flocculation basins and sedimentation basin occupy.

$$\text{Total area, ft}^2 = 4{,}500 \text{ ft}^2 + 20{,}150 \text{ ft}^2 = 24{,}650 \text{ ft}^2\text{, round to } \mathbf{25{,}000 \text{ ft}^2}$$

22. What is the area of a tank that has a radius of 14.5 ft?

First, find the diameter.

$$\textbf{Diameter, ft} = \textbf{(2)(Radius)}$$

$$\text{Diameter, ft} = (2)(14.5 \text{ ft}) \; 29.0 \text{ ft}$$

Next, determine the area.

The equation is:

$$\textbf{Area, ft}^2 = \textbf{(0.785)(Diameter)}^2$$

$$\text{Area, ft}^2 = (0.785)(29.0 \text{ ft})(29.0 \text{ ft}) = \mathbf{660 \text{ ft}^2}$$

23. What is the area of a triangular concrete slab that has a height of 12 ft and a base of 16 ft?

The equation is:

$$\textbf{Area, ft}^2 = \frac{\textbf{(Base, ft)(Height, ft)}}{\textbf{2}}$$

$$\text{Area, ft}^2 = \frac{(16\text{ ft})(12\text{ ft})}{2} = \mathbf{96\ ft^2}$$

24. If the area of a triangle is 24 ft² and the base is 8 ft, what is the height of the triangle?

The equation is:

$$\textbf{Area, ft}^2 = \frac{\textbf{(Base, ft)(Height, ft)}}{\textbf{2}}$$

Rearrange the equation and solve.

$$\text{Height, ft} = \frac{2(\text{Area, ft}^2)}{\text{Base, ft}}$$

$$\text{Height, ft} = \frac{2(24\text{ ft}^2)}{8\text{ ft}} = \mathbf{6\ ft}$$

25. What is the radius and area of a tank that has a circumference of 233 ft?

The equation is:

$$\textbf{Circumference, ft} = \mathbf{(\pi)(Diameter, ft)} \text{ or } \mathbf{2(\pi)(Radius, ft)}$$

Rearrange to solve for radius.

$$\text{Radius, ft} = \frac{\text{Circumference, ft}}{2(\pi)}$$

$$\text{Radius, ft} = \frac{233\text{ ft}}{2(3.14)} = \textbf{37.1 ft in radius}$$

(continued)

Next, find the area using the following equation:

$$\textbf{Area, ft}^2 = 2(\pi)r^2$$

$$\text{Area, ft}^2 = 2(3.14)(37.1\text{ ft})(37.1\text{ ft}) = 8{,}644\text{ ft}^2\text{, round to } \mathbf{8{,}640\text{ ft}^2}$$

26. What is the internal surface area of a cylindrical tank (bottom, top, and the cylinder wall) if it is 50.0 ft in diameter and 12.0 ft high? Assume the top is flat.

First, find the surface area of the tank's wall. Imagine cutting the wall and rolling it out so that it is flat. The length of this now flat wall is simply the circumference of the tank, and is defined as:

$$(\text{Diameter})(\pi)$$

The area of the tank wall then is the length of the wall times the height.

$$\text{Wall area, ft}^2 = (\text{Diameter})(\pi)(\text{Height})$$

Substituting:

$$\text{Wall area, ft}^2 = (50.0\text{ ft})(3.14)(12.0\text{ ft}) = 1{,}884\text{ ft}^2$$

Next, find the top and bottom surface area.

$$\text{Top and bottom area, ft}^2 = (2\text{, top and bottom})(0.785)(\text{Diameter})^2$$

$$\text{Top and bottom area, ft}^2 = (2)(0.785)(50.0\text{ ft})(50.0\text{ ft}) = 3{,}925\text{ ft}^2$$

$$\text{Total area of tank, ft}^2 = 1{,}884\text{ ft}^2 + 3{,}925\text{ ft}^2 = 5{,}809\text{ ft}^2\text{, round to } \mathbf{5{,}810\text{ ft}^2}$$

VOLUME DETERMINATIONS

27. Find the volume in gallons for a storage tank that is 18 ft in height and has a circumference of 215.8 ft.

First, find the diameter of the tank.

$$\textbf{Diameter, ft} = \frac{\textbf{Circumference, ft}}{\pi}$$

$$\text{Diameter, ft} = \frac{215.8\text{ ft}}{3.14} = 68.7\text{ ft}$$

Next, find the volume of the tank in gallons.

The equation is:

$$\textbf{Volume, gal} = \mathbf{(0.785)(Diameter)^2(Height)(7.48\ gal/ft^3)}$$

$$\text{Volume, gal} = (0.785)(68.7\text{ ft})(68.7\text{ ft})(18\text{ ft})(7.48\text{ gal/ft}^3)$$

$$= 498{,}835\text{ gal, round to } \mathbf{500{,}000\ gal}$$

28. A distribution pipe is 2.32 mi long. What is the volume of water in gallons if the pipe is 2.0 ft in diameter for a length of 1.75 mi and 18 in. for the remainder?

The equation is:

$$\textbf{Volume, gal} = \mathbf{(Cross\text{-}sectional\ area\ of\ the\ pipe,\ ft^2)(Length,\ ft)(7.48\ gal/ft^3)}$$

Therefore, we must first convert miles and inches to feet in order to use the above equation.

First, convert 1.75 mi to feet for the larger pipe.

$$(5{,}280\text{ ft/mi})(1.75\text{ mi}) = 9{,}240\text{ ft for larger pipe}$$

Then, determine the remainder of the feet for the smaller pipe.

$$2.32\text{ mi} - 1.75\text{ mi} = 0.57\text{ mi for smaller pipe}$$

Convert 0.57 mi to feet.

$$(5{,}280\text{ ft/mi})(0.57\text{ mi}) = 3{,}009.6\text{ ft}$$

Next, convert 18 in. to feet.

$$(18\text{ in.})(1\text{ ft/12 in.}) = 1.5\text{ ft}$$

Then, determine the cross-sectional area of each pipe size.

The equation is:

$$\textbf{Area, ft}^2 = \mathbf{(0.785)(Diameter)^2}$$

$$\text{Cross-sectional area for 2.0-ft diameter pipe, ft}^2 = (0.785)(2.0\text{ ft})(2.0\text{ ft}) = 3.14\text{ ft}^2$$

$$\text{Cross-sectional area for 1.5-ft diameter pipe, ft}^2 = (0.785)(1.5)(1.5) = 1.766\text{ ft}^2$$

(continued)

Volume, gal, for 2.0-ft diameter pipe = $(3.14 \text{ ft}^2)(9{,}240 \text{ ft})(7.48 \text{ gal/ft}^3)$ = 217,022 gal

Volume, gal, for 1.5-ft diameter pipe = $(1.766 \text{ ft}^2)(3{,}009.6 \text{ ft})(7.48 \text{ gal/ft}^3)$ = 39,756 gal

Last, add the two volumes together for the total volume.

Total volume, gal = 217,022 gal + 39,756 gal = 256,778 gal, round to **260,000 gal**

29. A tank is conical at the bottom and cylindrical at the top. If the diameter of the cylinder is 12.0 ft with a depth of 20.0 ft and the cone depth is 12.0 ft, what is the volume of the tank in cubic feet?

First, find the volume of the cone in cubic feet.

$$\textbf{Volume, ft}^3 = \tfrac{1}{3}\pi r^2\textbf{(Depth)}$$

where the radius = Diameter/2 = 12.0 ft/2 = 6.0 ft

$$\text{Volume, ft}^3 = \tfrac{1}{3}(3.14)(6.0 \text{ ft})(6.0 \text{ ft})(12.0 \text{ ft}) = 452 \text{ ft}^3$$

Next, find the volume of the cylindrical part of the tank.

$$\text{Volume} = \pi r^2(\text{Depth}) = (3.14)(6.0 \text{ ft})(6.0 \text{ ft})(20.0 \text{ ft}) = 2{,}261 \text{ ft}^3$$

Last, add the two volumes for the answer.

$$\text{Total volume, ft}^3 = 452 \text{ ft}^3 + 2{,}261 \text{ ft}^3 = 2{,}713 \text{ ft}^3\text{, round to } \mathbf{2{,}710 \text{ ft}^3}$$

30. A storage tank is 110.0 ft in diameter and has an overflow 34.5 ft above the base of the tank. How many gallons of water are in the tank if it is 72.4% full?

First, determine the maximum volume in gallons the tank can hold.

The equation is:

$$\textbf{Volume, gal} = \textbf{(0.785)(Diameter)}^2\textbf{(Depth or height, ft)(7.48 gal/ft}^3\textbf{)}$$

$$\text{Volume, gal} = (0.785)(110.0\text{ ft})(110.0\text{ ft})(34.5\text{ ft})(7.48\text{ gal/ft}^3) = 2{,}451{,}183$$

Now, determine the number of gallons in the tank if it is 72.4% full.

$$\text{Number of gal} = \frac{(2{,}451{,}183\text{ gal})(72.4\%)}{100\%} = 1{,}774{,}656\text{ gal, round to } \mathbf{1{,}770{,}000\text{ gal}}$$

31. A trench that is 3.5 ft wide, 980 ft in length, and 7.0 ft in depth is excavated for the purpose of installing a water main. Determine the number of cubic feet and cubic yards that need to be excavated for this water main.

First, determine the number of cubic feet excavated from the trench.

The equation is:

$$\textbf{Volume, ft}^3 = \textbf{(Length, ft)(Width, ft)(Depth, ft)}$$

$$\text{Volume, ft}^3 = (980\text{ ft})(3.5\text{ ft})(7.0\text{ ft}) = 24{,}010\text{ ft}^3\text{, round to } \mathbf{24{,}000\text{ ft}^3}$$

Next, find the number of cubic yards.

$$\text{Number of yd}^3 = \frac{24{,}010\text{ ft}^3}{27\text{ ft}^3/\text{yd}^3} = 889.26\text{ yd}^3\text{, round to } \mathbf{890\text{ yd}^3}$$

32. Find the volume in cubic feet and gallons of a trapezoidal trough that has the following dimensions:

Length = 25.0 ft	**Height = 3.5 ft**
Bottom width = 4.0 ft (b_1)	**Top width = 7.5 ft (b_2)**

The equation is:

$$\textbf{Volume, ft}^3 = \frac{(b_1 + b_2)(\text{Height})(\text{Length})}{2}$$

$$\text{Volume, ft}^3 = \frac{(4.0 \text{ ft} + 7.5 \text{ ft})(3.5 \text{ ft})(25.0 \text{ ft})}{2}$$

$$\text{Volume, ft}^3 = \frac{(11.5 \text{ ft})(3.5 \text{ ft})(25.0 \text{ ft})}{2} = \frac{1{,}006.25}{2} = 503.125 \text{ ft}^3\text{, round to } \mathbf{500\ ft^3}$$

Next, find the number of gallons.

Number of gal = (503.125 ft^3)(7.48 gal/ft^3) = 3,763.4 gal, round to **3,800 gal**

33. What is the volume in gallons for a canal that is 2,375 ft in length, has a semi-circular shape (half circle) with a diameter of 8.0 ft, and a water depth of 3.5 ft?

First, find the area of the half-circle.

The equation is:

$$\textbf{Area, ft}^2 = \frac{(0.785)(\text{Diameter})^2}{2}$$

$$\text{Area, ft}^2 = \frac{(0.785)(8.0 \text{ ft})(8.0 \text{ ft})}{2} = 25.12 \text{ ft}^2$$

Next, find the volume in gallons.

Volume, gal = (Area, ft^2)(Length of canal, ft)(7.48 gal/ft^3)

Volume, gal = (25.12 ft^2)(2,375, ft)(7.48 gal/ft^3) = 446,257 gal, round to **450,000 gal**

34. A lime silo is conical at the bottom and cylindrical at the top. If the diameter of the cylinder is 15.0 ft with a depth of 30.0 ft, and the cone depth is 10.5 ft, what is the volume of the tank in cubic feet?

First, find the volume of the cone in cubic feet.

$$\textbf{Volume, ft}^3 = \tfrac{1}{3}\pi r^2\textbf{(Depth)}$$

where the radius = Diameter/2 = 15.0 ft/2 = 7.50 ft

$$\text{Volume, ft}^3 = \tfrac{1}{3}(3.14)(7.5 \text{ ft})(7.5 \text{ ft})(10.5 \text{ ft}) = 618.19 \text{ ft}^3$$

Next, find the volume of the cylindrical part of the tank.

$$\text{Volume} = \pi r^2(\text{Depth}) = (3.14)(7.5 \text{ ft})(7.5 \text{ ft})(30.0 \text{ ft}) = 5{,}298.75 \text{ ft}^3$$

Last, add the two volumes for the answer.

Total volume, ft^3 = 618.19 ft^3 + 5,298.75 ft^3 = 5,916.94 ft^3, round to **5,920 ft^3**

35. A tank is conical at the bottom and cylindrical at the top. If the diameter of the cylinder is 25.0 ft with a depth of 38.0 ft and the cone depth is 13.5 ft, what is the volume of the tank in cubic feet and gallons?

First, find the volume of the cone in cubic feet

$$\textbf{Volume, ft}^3 = \tfrac{1}{3}\pi r^2\textbf{(Depth)}$$

where the radius = Diameter/2 = 25.0 ft/2 = 12.5 ft

$$\text{Volume, ft}^3 = \tfrac{1}{3}(3.14)(12.5 \text{ ft})(12.5 \text{ ft})(13.5 \text{ ft}) = 2{,}207.81 \text{ ft}^3$$

Next, find the volume of the cylindrical part of the tank.

$$\text{Volume} = \pi r^2(\text{depth}) = (3.14)(12.5 \text{ ft})(12.5 \text{ ft})(38.0 \text{ ft}) = 18{,}643.75 \text{ ft}^3$$

Then, add the two volumes for the answer.

Total volume, ft^3 = 2,207.81 ft^3 + 18,643.75 ft^3 = 20,852 ft^3, round to **20,900 ft^3**

To find the number of gallons, multiply the total number of cubic feet by 7.48 gal/ft^3.

(20,852 ft^3)(7.48 gal/ft^3) = 155,973 gal, round to **156,000 gal**

AVERAGE CALCULATIONS

36. What is the average in pounds per day for chlorine used given the following data?

Mon.	Tues.	Wed.	Thurs.	Fri.	Sat.	Sun.
209	218	225	257	265	268	289

The equation is:

$$\textbf{Average } \mathbf{Cl_2} \textbf{ used, lb/d} = \frac{\textbf{Sum of } \mathbf{Cl_2} \textbf{ used each day, lb}}{\textbf{Total time, d}}$$

$$\text{Average } Cl_2 \text{ used, lb/d} = \frac{209 + 218 + 225 + 257 + 265 + 268 + 289}{7 \text{ days}} =$$

$$\frac{1{,}731 \text{ lb}}{7 \text{ d}} = \mathbf{247 \textbf{ lb/d of } Cl_2}$$

37. What is the average million gallons per day flow from a storage tank given the following data?

Mon.	Tues.	Wed.	Thurs.	Fri.	Sat.	Sun.
4.2	3.8	3.9	3.6	3.5	3.5	3.3

The equation is:

$$\frac{\textbf{Average, million gallons per day flow}}{\textbf{Total time, d}} = \textbf{Sum of mgd used each day}$$

$$\text{Average, million gallons per day flow} =$$

$$\frac{4.2 + 3.8 + 3.9 + 3.6 + 3.5 + 3.5 + 3.3}{7 \text{ d}} = 3.69 \text{ mgd, round to } \mathbf{3.7 \textbf{ mgd}}$$

38. The average water production from a water plant is given below for each week in the month of June. What is the average water production for the treatment plant in million gallons per day?

First, week	Second week	Third week	Fourth week	Fifth week (last two days)
48.35	49.22	49.83	50.11	15.02

The equation is:

$$\frac{\textbf{Average, million gallons per day produced}}{\textbf{Total time, d}} = \textbf{Sum of mgd used each week}$$

$$\text{Average, million gallons per day produced} = \frac{48.35 + 49.22 + 49.83 + 50.11 + 15.02}{30\text{ d}} = \mathbf{7.084\ mgd}$$

DETENTION TIME CALCULATIONS

39. Calculate the theoretical detention time in hours for the following basins of a treatment plant:

Four flocculation (floc) basins, each 60.0 ft by 20.2 ft with a water depth of 9.5 ft

A sedimentation (sed) basin that is 345 ft long, 78.5 ft wide, and has a water depth of 12.5 ft. The flow is 21.2 mgd.

Determine the number of gallons in the four flocculation basins and the sed basin.

The equation is

$$\textbf{Volume, gal} = \textbf{(Length)(Width)(Depth)(7.48 gal/ft}^3\textbf{)(Number of basins)}$$

Volume, gal, in floc basins = (60.0 ft)(20.2 ft)(9.5 ft)(7.48 gal/ft^3)(4 basins) = 344,499 gal

Volume, gal, in sed basin = (345 ft)(78.5 ft)(12.5 ft)(7.48 gal/ft^3) = 2,532,214 gal

2,876,713 gal

Next, convert million gallons per day to gallons per hour.

$$(21.2\text{ mgd})(1\text{ d}/24\text{ hr})(1{,}000{,}000/1\text{ M}) = 883{,}333\text{ gph}$$

(continued)

Write the equation with units asked for in the question.

$$\textbf{Detention time, hr} = \frac{\textbf{Volume, gal}}{\textbf{Flow rate, gph}}$$

$$\text{Detention time, hr} = \frac{2{,}876{,}713 \text{ gal}}{883{,}333 \text{ gph}} = \textbf{3.26 hr}$$

40. Calculate the theoretical detention time in hours for the following water treatment plant's clear well and associated distribution line and storage tank:

The clear well is 675 ft long, 125 ft wide, and has an average water depth of 16.25 ft. The distribution line from clear well to storage tank is 1.57 mi long and has a diameter of 2.5 ft. The storage tank has a diameter of 80.0 ft and averages a height of 24.4 ft of water. The flow is 44.1 mgd.

Determine the number of gallons in each of the containment structures.

The equation is:

$$\textbf{Volume, gal} = \textbf{(Length)(Width)(Depth)(7.48 gal/ft}^3\textbf{)}$$

$$\text{Volume, gal, in clear well} = (675 \text{ ft})(125 \text{ ft})(16.25 \text{ ft})(7.48 \text{ gal/ft}^3) = 10{,}255{,}781 \text{ gal}$$

Next, convert number of miles to feet.

$$\text{Number of ft} = (5{,}280 \text{ ft/mi})(1.57 \text{ mi}) = 8{,}289.6 \text{ ft}$$

The equation is:

$$\textbf{Volume, gal} = \textbf{(Length)(0.785)(Diameter)}^2\textbf{(7.48 gal/ft}^3\textbf{)}$$

$$\text{Volume, gal, in pipeline} = (8{,}289.6 \text{ ft})(0.785 \text{ ft})(2.5 \text{ ft})(2.5 \text{ ft})(7.48 \text{ gal/ft}^3) = 304{,}218 \text{ gal}$$

$$\text{Volume, gal, in tank} = (0.785)(80.0 \text{ ft})(80.0 \text{ ft})(24.4 \text{ ft})(7.48 \text{ gal/ft}^3) = 916{,}940 \text{ gal}$$

$$11{,}476{,}939 \text{ gal}$$

Next, convert million gallons per day to gallons per hour.

$$(44.1 \text{ mgd})(1 \text{ d/24 hr})(1{,}000{,}000/1 \text{ M}) = 1{,}837{,}500 \text{ gph}$$

Write the equation with units asked for in the question.

$$\textbf{Detention time, hr} = \frac{\textbf{Volume, gal}}{\textbf{Flow rate, gph}}$$

$$\text{Detention time, hr} = \frac{11{,}476{,}939 \text{ gal}}{1{,}837{,}500 \text{ gph}} = 6.246 \text{ hr, round to } \mathbf{6.2\ hr}$$

41. Calculate the theoretical detention time in hours for the following:

The clear well is 550 ft long, 115 ft wide, and has an average water depth of 13.25 ft. The distribution pipe from the clear well to the storage tank is 2.21 mi long and has a diameter of 2.0 ft. The storage tank has a diameter of 60.0 ft and averages a height of 23.25 ft of water. The flow is 31.7 mgd.

Determine the number of gallons in the clear well, distribution pipe, and storage tank.

The equation is:

$$\textbf{Volume, gal} = \textbf{(Length)(Width)(Depth)(7.48 gal/ft}^3\textbf{)(Number of basins)}$$

$$\text{Volume, gal, in clear well} = (550 \text{ ft})(115 \text{ ft})(13.25 \text{ ft})(7.48 \text{ gal/ft}^3) = 6{,}268{,}708 \text{ gal}$$

Next, convert number of miles to feet.

$$\text{Number of feet} = (5{,}280 \text{ ft/mi})(2.21 \text{ mi}) = 11{,}668.8 \text{ ft}$$

$$\textbf{Volume, gal, in pipeline} = \textbf{(Length)(0.785)(Diameter)}^2\textbf{(7.48 gal/ft}^3\textbf{)}$$

$$\text{Volume, gal, in pipeline} = (11{,}668.8 \text{ ft})(0.785 \text{ ft})(2.0 \text{ ft})(2.0 \text{ ft})(7.48 \text{ gal/ft}^3) = 274{,}067 \text{ gal}$$

$$\textbf{Volume, gal, in storage tank} = \textbf{(0.785)(Diameter)}^2\textbf{(Height)(7.48 gal/ft}^3\textbf{)}$$

$$\text{Volume, gal, in storage tank} = (0.785)(60.0 \text{ ft})(60.0 \text{ ft})(23.25 \text{ ft})(7.48 \text{ gal/ft}^3) = 491{,}470 \text{ gal}$$

$$7{,}034{,}245 \text{ gal}$$

Next, convert million gallons per day to gallons per hour.

$$(31.7 \text{ mgd})(1 \text{ d}/24 \text{ hr})(1{,}000{,}000/1 \text{ M}) = 1{,}320{,}833 \text{ gph}$$

Write the equation with units asked for in the question.

$$\textbf{Detention time, hr} = \frac{\textbf{Volume, gal}}{\textbf{Flow rate, gph}}$$

$$\text{Detention time, hr} = \frac{7{,}034{,}245 \text{ gal}}{1{,}320{,}833 \text{ gph}} = 5.33 \text{ hr, round to } \mathbf{5.3\ hr}$$

WATERSHED SUPPLY CALCULATIONS

42. A water treatment plant distributes 5,610 mil gal/yr. Given the following data, how many million gallons are available in the watershed in a typical year, what is the average water production per day, and does this water treatment plant have to import water to meet the water demand?

Average annual rainfall is 9.91 in.

The watershed is 299 mi^2.

7.54% of the rainfall is collected for treatment.

First, find the number of inches collected from the rainfall.

$$(9.91 \text{ in.})(7.54\%/100\%) = 0.747 \text{ in.}$$

Convert the number of inches of rain collected to feet.

$$(0.747 \text{ in.})(1 \text{ ft}/12 \text{ in.}) = 0.062 \text{ ft}$$

Then, calculate the number of gallons in 1 acre that is covered by 0.062 ft of water.

$$(43{,}560 \text{ ft}^2/\text{acre})(7.48 \text{ gal/ft}^3)(0.062 \text{ ft}) = 20{,}201 \text{ gal/acre}$$

Convert this to million gallons.

$$(20{,}201 \text{ gal/acre})(1 \text{ M}/1{,}000{,}000) = 0.0202 \text{ mil gal/acre}$$

Next, find the number of acres in 299 mi^2.

$$\text{Number of acres} = (640 \text{ acres/mi}^2)(299 \text{ mi}^2) = 191{,}360 \text{ acres}$$

Then, calculate the number of million gallons collected from the watershed.

(0.0202 mil gal/acre)(191,360 acres) = 3,865.5 mil gal/yr, round to **3,870 mil gal/yr**

Yes, the water treatment plant needs to import water as they require 5,610 mil gal/yr and have only 3,870 mil gal/yr available.

43. Calculate the rainfall for a watershed given the following data and assuming the treatment plant collects and treats 8.4% of the water:

The treatment facility collects 4,874 mil gal for the year.

The watershed is 244 mi^2.

First, calculate the number of acres in 244 mi^2.

$$(640 \text{ acres/mi}^2)(244 \text{ mi}^2) = 156{,}160 \text{ acres}$$

Next, determine the number of gallons per acre.

$$\frac{(4{,}874 \text{ mil gal})(1{,}000{,}000/1 \text{ M})}{156{,}160 \text{ acres}} = 31{,}212 \text{ gal/acre}$$

Because this is only 8.4% of the water, we need to divide by 8.4% and multiply by 100% to get the actual rainfall per acre.

$$\frac{(31{,}212 \text{ gal/acre})(100\%)}{8.4\%} = 371{,}571 \text{ gal/acre}$$

Then, find the thickness this water would have over the acre in inches.

$$\frac{(371{,}571 \text{ gal/acre})(12 \text{ in./ft})}{(43{,}560 \text{ ft}^2\text{/acre})(7.48 \text{ gal/ft}^3)} = 13.68 \text{ in., round to } \mathbf{14 \text{ in.}}$$

PRESSURE PROBLEMS

44. If the water level in a tank is 31.78 ft, what is the pressure in psi at the bottom?

The equation is:

$$\mathbf{psi} = \frac{\mathbf{Depth, ft}}{\mathbf{2.31 \text{ ft/psi}}}$$

$$\text{psi} = \frac{31.78 \text{ ft}}{2.31 \text{ ft/psi}} = \mathbf{13.8 \text{ psi}}$$

45. If the pressure at the bottom of the tank is 14.7 psi, what is the height of the water in the tank?

The equation is:

$$\textbf{Depth, ft} = \textbf{(Number of psi)(2.31 ft/psi)}$$

$$\text{Water level in tank} = (14.7 \text{ psi})(2.31 \text{ ft/psi}) = \mathbf{34.0\ ft}$$

46. If the pressure head at a fire hydrant is 210 ft, what is the psi?

The equation is:

$$\textbf{Pressure head, ft} = \textbf{(Pressure, psi)(2.31 ft/psi)}$$

Rearrange the equation to solve for pressure.

$$\text{Pressure, psi} = \frac{\text{Pressure head, ft}}{2.31 \text{ ft/psi}}$$

$$\text{Pressure, psi} = \frac{210 \text{ ft}}{2.31 \text{ ft/psi}} = \mathbf{91\ psi}$$

FLOW PROBLEMS

47. If flow through a 10.0-in. pipe is 563 gpm, what must the velocity be in feet per second?

First, convert the number of gallons per minute to cubic feet per second.

$$\text{Number of ft}^3\text{/s} = \frac{563 \text{ gpm}}{(7.48 \text{ gal/ft}^3)(60 \text{ s/min})} = 1.25 \text{ ft}^3\text{/s}$$

Next, convert the diameter from inches to feet.

$$\text{Number of ft} = (10 \text{ in.})(1 \text{ ft/12 in.}) = 0.833 \text{ ft}$$

Equation for flow is:

$$\textbf{Flow, ft}^3\textbf{/s} = \textbf{(Area, ft}^2\textbf{)(Velocity, ft/s)}$$

where the area = (0.785)(Diameter)^2

$$1.61 \text{ ft}^3\text{/s} = (0.785)(0.833 \text{ ft})(0.833 \text{ ft})(\text{Flow, ft/s})$$

Rearrange the equation to solve for flow.

$$\text{Flow, ft/s} = \frac{1.25 \text{ ft}^3\text{/s}}{(0.785)(0.833 \text{ ft})(0.833 \text{ ft})} = \mathbf{2.29 \text{ ft/s}}$$

48. An 8.0-in. diameter distribution pipe delivers 1,011,000 gal in 24 hr. What is the average flow during the 24-hr time period in feet per second?

First, convert the diameter in inches to feet.

$$\text{Number of feet} = (8.0 \text{ in.})(1 \text{ ft}/12 \text{ in.}) = 0.667$$

Next, determine the cross-sectional area of the pipe.

The equation is:

$$\mathbf{Area, ft^2 = (0.785)(Diameter)^2}$$

$$\text{Area, ft}^2 = (0.785)(0.667 \text{ ft})(0.667 \text{ ft}) = 0.349 \text{ ft}$$

Next, determine the cubic feet per second.

$$\text{Number of ft}^3\text{/s} = (1{,}011{,}000 \text{ gal}/24 \text{ hr})(1 \text{ ft}^3/7.48 \text{ gal})(1 \text{ hr}/3{,}600 \text{ s}) = 1.56 \text{ ft}^3\text{/s}$$

Last, determine the flow in feet per second.

The equation is:

$$\mathbf{Flow, ft^3/s = (Area, ft^2)(Velocity, ft/s)}$$

Rearrange the formula to solve for velocity.

$$\text{Velocity, ft/s} = \frac{\text{Flow, ft}^3\text{/s}}{\text{Area, ft}^2}$$

Substituting:

$$\text{Velocity, ft/s} = \frac{1.56 \text{ ft}^3\text{/s}}{0.349 \text{ ft}^2} = 4.47 \text{ ft/s, round to } \mathbf{4.5 \text{ ft/s}}$$

49. What is the flow through a 6.0-in. diameter pipe if it delivers 875 gpm?

First, convert the diameter of the pipe to feet.

$$(6.0 \text{ in.})(1 \text{ ft}/12 \text{ in.}) = 0.5 \text{ ft}$$

Next, determine the cross-sectional area of the pipe.

The equation is:

$$\textbf{Area, ft}^2 = \textbf{(0.785)(Diameter)}^2$$

$$\text{Area, ft}^2 = (0.785)(0.5 \text{ ft})(0.5 \text{ ft}) = 0.196 \text{ ft}^2$$

Then, convert gallons per minute to cubic feet per second.

$$\text{Number of ft}^3\text{/s} = \frac{875 \text{ gpm}}{(7.48 \text{ gal/ft}^3)(60 \text{ s/min})} = 1.95 \text{ ft}^3\text{/s}$$

Last, determine the flow in feet per second.

The equation is:

$$\textbf{Flow, ft/s} = \textbf{(Area, ft}^2\textbf{)(Flow, ft}^3\textbf{/s)}$$

$$\text{Flow, ft/s} = (0.196 \text{ ft}^2)(1.95 \text{ ft}^3\text{/s}) = 0.382 \text{ ft/s, round to } \textbf{0.38 ft/s}$$

50. A water channel is 8.25 ft wide and averages 3.75 ft in depth. What is the velocity of the water if the flow is 45 ft³/s?

The equation is:

$$\textbf{Flow, ft}^3\textbf{/s} = \textbf{(Area, ft}^2\textbf{)(Velocity, ft/s)}$$

commonly written as:

$$\mathbf{Q = AV}$$

Rearrange the equation to solve for velocity.

$$\text{Velocity, ft/s} = \frac{\text{Flow, ft}^3\text{/s}}{\text{Area, ft}^2}$$

$$\text{Velocity, ft/s} = \frac{45 \text{ ft}^3\text{/s}}{(8.25 \text{ ft})(3.75 \text{ ft})} = 1.45 \text{ ft/s, round to } \textbf{1.4 ft/s}$$

51. The velocity through a channel is 2.10 ft/s. If the channel is 6.5 ft wide and 2.8 ft in depth, what is the flow in cubic feet per second?

The equation is:

$$\mathbf{Q = AV, \text{ or Flow, ft}^3\text{/s} = (\text{Area, ft}^2)(\text{Velocity, ft/s})}$$

$$\text{Q, ft}^3\text{/s} = (2.10 \text{ ft/s})(6.5 \text{ ft})(2.8 \text{ ft}) = 38.22 \text{ ft}^3\text{/s, round to } \mathbf{38 \text{ ft}^3\text{/s}}$$

52. Water is flowing through a faucet at 15.5 gpm. How long will it take to fill a swimming pool in hours and minutes if the pool is 45 ft by 22 ft and averages 5.5 ft in depth?

First, determine the capacity of the swimming pool in gallons.

The equation is:

$$\mathbf{\text{Volume, gal} = (\text{Length})(\text{Width})(\text{Depth})(7.48 \text{ gal/ft}^3)}$$

$$\text{Volume, gal} = (45 \text{ ft})(22 \text{ ft})(5.5 \text{ ft})(7.48 \text{ gal/ft}^3) = 40{,}728.6 \text{ gal}$$

Next, find the time to fill the pool.

The equation is:

$$\mathbf{\text{Time, min} = \frac{\text{Volume, gal}}{\text{Rate, gpm}}}$$

$$\text{Time, min} = \frac{40{,}728.6 \text{ gal}}{15.5 \text{ gpm}} = 2{,}627.65 \text{ min}$$

Last, convert minutes to hours.

$$\text{Time, hr} = \frac{2{,}627.65 \text{ min}}{60 \text{ min/hr}} = 43.79 \text{ hr} = \mathbf{43 \text{ hr and } 47 \text{ min}}$$

53. If a 5.0-gal bucket is filled in 17 s, what is the flow from the faucet in gallons per minute?

First, convert the number of seconds to minutes.

$$\text{Time, min} = \frac{17 \text{ s}}{60 \text{ s/min}} = 0.283 \text{ min}$$

Next, determine the flow in gallons per minute.

The equation is:

$$\textbf{Flow, gpm} = \textbf{Volume, gal/Time, min}$$

$$\text{Flow, gpm} = \frac{5.0 \text{ gal}}{0.283 \text{ min}} = 17.67 \text{ gpm, round to } \textbf{18 gpm}$$

FLUSHING TIME PROBLEMS

54. An 8.0-in. main line needs to be flushed. If the desired length of pipeline to be flushed is 250 ft, how many minutes will it take to flush the line at 25 gpm?

First, convert the pipe diameter from inches to feet.

$$\text{Number of ft} = (8.0 \text{ in.})(1 \text{ ft/12 in.}) = 0.667 \text{ ft}$$

Next, determine the volume in gallons for 250 ft of the 8-in. pipeline.

The equation is:

$$\textbf{Volume, gal} = \textbf{(0.785)(Diameter)}^2\textbf{(Length)(7.48 gal/ft}^3\textbf{)}$$

$$\text{Volume, gal} = (0.785)(0.667 \text{ ft})(0.667 \text{ ft})(250 \text{ ft})(7.48 \text{ gal/ft}^3) = 653.07 \text{ gal}$$

Then, find the flushing time.

$$\text{Flushing time, min} = \frac{653.07 \text{ gal}}{25 \text{ gpm}} = 26.1 \text{ min, round to } \textbf{26 min}$$

55. A 22.5-ft, ¾-in. service line requires flushing. How many minutes are required to flush the line 25 times if the line is flushed at a rate of 8.0 gpm?

First, convert the pipe diameter from inches to feet.

Number of ft = (¾ in.)(1 ft/12 in.) = (0.75 in.)(1 ft/12 in.) = 0.0625 ft

Next, determine the volume in gallons for 22.5 ft of the ¾-in. pipeline.

The equation is:

$$\textbf{Volume, gal} = \textbf{(0.785)(Diameter)}^2\textbf{(Length)(7.48 gal/ft}^3\textbf{)(25 volumes)}$$

$$\text{Volume, gal} = (0.785)(0.0625 \text{ ft})(0.0625 \text{ ft})(22.5 \text{ ft})(7.48 \text{ gal/ft}^3)(25 \text{ volumes}) = 12.9 \text{ gal}$$

Then, find the flushing time.

$$\text{Flushing time, min} = \frac{12.9 \text{ gal}}{8.0 \text{ gpm}} = 1.61 \text{ min, round to } \textbf{1.6 min}$$

DENSITY CALCULATIONS

56. What is the density of a substance that weighs 401.8 g and occupies a space of 389.1 cm³?

The equation is:

$$\textbf{Density} = \textbf{Mass/Volume}$$

$$\text{Density} = 401.8 \text{ g}/389.1 \text{ cm}^3 = \textbf{1.033 g/cm}^3$$

57. A liquid alum solution weighs 28.25 lb and occupies 11.185 L. What is the density of the alum in grams per cubic centimeter?

First, convert the number of pounds to grams.

The following applies:

$$1 \text{ lb} = 454 \text{ g and } 1 \text{ L} = 1000.027 \text{ cm}^3$$

Number of g = (Number of lb)(454 g/lb)

Number of g = (28.25 lb)(454 g/lb) = 12,825.5 g

Number of cm^3 = (1000.027 cm^3/1 L)(11.185 L) = 11,185 cm^3

Density = Mass/Volume

Density = 12,825 g/11,185 cm^3 = **1.15 g/cm^3**

58. A polymer solution weighs 2.14 lb and occupies a space of exactly 750 mL. What is its density in pounds per gallon?

First, convert the number of milliliters to gallons.

Number of gal = (Number of mL)(1 gal/3,785 mL)

Number of gal = (750 mL)(1 gal/3,785 mL) = 0.198 gal

Density = Mass/Volume

In this case, weight is substituted for mass.

Density = 2.14 lb/0.198 gal = **10.8 lb/gal**

59. How many pounds per gallon will a solution weigh if it has a specific gravity (sp gr) of 1.09?

The equation is:

Specific gravity = Density of substance/Density of water

Rearrange the equation to solve for the density of the liquid in pounds per cubic foot.

Density of liquid, lb/ft^3 = (Specific gravity)(Density of water, lb/gal)

Density of liquid, lb/ft^3 = (1.09)(8.34 lb/gal) = **9.09 lb/gal**

SPECIFIC GRAVITY OF LIQUIDS

Specific gravity (sp gr) is a dimensionless number that compares the density of one substance to that of another. Water is the standard for liquids with a specific gravity of 1.0. The density of water can be expressed in several ways: 62.4 lb/ft^3, 8.34 lb/gal, or 1 g/cm^3.

60. What is the specific gravity of a polymer solution that weighs 11.1 lb/gal?

First, look at the units so that you can compare the density of water to the unknown in similar units. In this case, we know that water has a density of 8.34 lb/gal. Divide the density of the unknown by the density of water in pounds per gallon.

The equation is:

Specific gravity = Density of unknown substance/Density of water

$$\text{Specific gravity of polymer solution} = \frac{11.1 \text{ lb/gal}}{8.34 \text{ lb/gal}} = \mathbf{1.33 \text{ sp gr}}$$

61. Determine the specific gravity of a gold bar that weighs 521.47 lb and occupies a space of 0.433 ft^3.

First, determine the number of pounds of gold you would have if it occupied 1 ft^3.

$$\text{Number of lb/ft}^3 = \frac{521.47 \text{ lb}}{0.433 \text{ ft}^3} = 1{,}204.32 \text{ lb/ft}^3$$

To find the specific gravity, choose 62.4 lb/ft^3 for comparing water density to the polymer density because they have similar units.

The equation is:

Specific gravity = Density of unknown substance/Density of water

$$\text{Specific gravity for gold} = \frac{1{,}204.32 \text{ lb/ft}^3}{62.4 \text{ lb/ft}^3} = \mathbf{19.3 \text{ sp gr}}$$

METER ACCURACY CALCULATIONS

62. A meter being tested by a laboratory shows a reading of 2,255 gal. A volumetric tank used to measure the water that flowed through the meter indicates the actual volume is 2,271 gal. What is the percent accuracy of the meter?

The equation is:

$$\textbf{Meter accuracy, \%} = \frac{\textbf{(Meter reading, gal)(100\%)}}{\textbf{Actual volume, gal}}$$

$$\text{Meter accuracy, \%} = \frac{(2{,}255 \text{ gal})(100\%)}{2{,}271 \text{ gal}} = \textbf{99.30\% meter accuracy}$$

63. A meter being tested reads 1,982 L. A volumetric tank shows the actual value is 2,016 L. What is the percent accuracy of the meter?

The equation is:

$$\textbf{Meter accuracy, \%} = \frac{\textbf{(Meter reading, gal)(100\%)}}{\textbf{Actual volume, gal}}$$

$$\text{Meter accuracy, \%} = \frac{(1{,}982 \text{ L})(100\%)}{2{,}016 \text{ L}} = \textbf{98.31\% meter accuracy}$$

RATIO PROBLEMS

64. Twenty-five pounds of soda ash are dissolved in 92 gal of water to produce a desired concentration. How many pounds of soda ash are needed to make the same desired concentration if 30 gal of water are used?

This type of problem, where there is one unknown, can be set up as a ratio.

Let x = unknown (pounds of soda ash).

The equation is:

$$\frac{\textbf{\textit{x} pounds, soda ash}}{\textbf{Number of gal}_2} = \frac{\textbf{Number of pounds, soda ash}}{\textbf{Number of gal}_1}$$

$$\frac{x \text{ pounds, soda ash}}{30 \text{ gal}} = \frac{25 \text{ lb, soda ash}}{92 \text{ gal}}$$

$$x \text{ pounds, soda ash} = \frac{(25 \text{ lb, soda ash})(30 \text{ gal})}{92 \text{ gal}} = 8.15 \text{ lb, round to } \mathbf{8 \text{ lb of soda ash}}$$

65. The alum dosage for a water treatment facility is 478 mL/min at a flow rate of 12.5 ft^3/s. If the flow is increased to 18 ft^3/s, what will be the theoretical alum dosage in milliliters per minute if everything else remains the same?

Let x = unknown (mL/min, alum).

The equation is:

$$\frac{\mathbf{x \text{ mL/min, alum}}}{\mathbf{\text{Flow rate}_2, \text{ft}^3/\text{s}}} = \frac{\mathbf{\text{Number of mL/min, alum}}}{\mathbf{\text{Flow rate}_1, \text{ft}^3/\text{s}}}$$

$$\frac{x \text{ mL/min, alum}}{18 \text{ ft}^3/\text{s}} = \frac{478 \text{ mL/min, alum}}{12.5 \text{ ft}^3/\text{s}}$$

$$x \text{ mL/min, alum} = \frac{(478 \text{ mL/min, alum})(18 \text{ ft}^3/\text{s})}{12.5 \text{ ft}^3/\text{s}}$$

$$x \text{ mL/min, alum} = 688.3 \text{ mL/min, round to } \mathbf{690 \text{ mL/min, alum}}$$

66. A raw water flow for a treatment plant of 44 ft^3/s is being prechlorinated with chlorine gas at a rate of 370 lb/d. If the flow is changed to 28 ft^3/s, what should the new chlorinator setting be in pounds per day?

Let x = unknown (pounds per day, Cl_2).

The equation is:

$$\frac{\mathbf{x \text{ lb/d, Cl}_2}}{\mathbf{\text{Flow rate}_2, \text{ft}^3/\text{s}}} = \frac{\mathbf{\text{Number of lb/d, Cl}_2}}{\mathbf{\text{Flow rate}_1, \text{ft}^3/\text{s}}}$$

$$\frac{x \text{ lb/d, Cl}_2}{28 \text{ ft}^3/\text{s}} = \frac{370 \text{ lb/d Cl}_2}{44 \text{ ft}^3/\text{s}}$$

$$x \text{ lb/d, Cl}_2 = \frac{(28 \text{ ft}^3/\text{s})(370 \text{ lb/d Cl}_2)}{44 \text{ ft}^3/\text{s}} = 235.45 \text{ lb/d, round to } \mathbf{235 \text{ lb/d of Cl}_2}$$

CALCULATIONS FOR CHLORINE DOSE, DEMAND, AND RESIDUAL

67. If the chlorine dose is 3.85 mg/L and the chlorine residual is 1.45 mg/L, what is the chlorine demand?

The equation is:

Chlorine dose = Chlorine demand + Chlorine residual

Rearrange and solve for chlorine demand.

Chlorine demand = Chlorine dose – Chlorine residual

Chlorine demand = 3.85 mg/L – 1.45 mg/L = **2.40 mg/L**

68. If the chlorine dose is 3.25 mg/L and the chlorine demand is 1.88 mg/L, what is the chlorine residual?

The equation is:

Chlorine dose = Chlorine demand + Chlorine residual

Rearrange and solve for chlorine residual.

Chlorine residual = Chlorine dose – Chlorine demand

Chlorine residual = 3.25 mg/L – 1.88 mg/L = **1.37 mg/L**

69. If the chlorine residual is 1.14 mg/L and the chlorine demand is 3.01 mg/L, what is the chlorine dose?

The equation is:

Chlorine dose = Chlorine demand + Chlorine residual

Chlorine dose = 1.14 mg/L + 3.01 mg/L = **4.15 mg/L**

DOSAGE PROBLEMS

70. A storage tank requires disinfection with a 65% calcium hypochlorite solution. The tank is 35.0 ft in diameter and the overflow is 17.25 ft above the bottom of the tank. How many gallons of hypochlorite solution are required if the desired dosage is 50.0 mg/L? Assume the calcium hypochlorite solution weighs 10.21 lb/gal.

First, calculate the volume in gallons for the storage tank.

$$\textbf{Volume, gal} = \mathbf{(0.785)(Diameter)^2(Depth\ or\ height)(7.48\ gal/\ ft^3)}$$

$$\text{Volume, gal} = (0.785)(35.0\text{ ft})(35.0\text{ ft})(17.25\text{ ft})(7.48\text{ gal/ft}^3) = 124{,}078\text{ gal}$$

Next, convert gallons to million gallons.

$$\text{Number of mil gal} = (124{,}078\text{ gal})(1\text{ M}/1{,}000{,}000) = 0.124\text{ mil gal}$$

Next, determine the number of pounds of chlorine required.

$$\text{Number of lb} = (\text{Number of mil gal})(\text{Dosage, mg/L})(8.34\text{ lb/gal})$$

$$\text{Number of lb} = (0.124\text{ mil gal})(50.0\text{ mg/L})(8.34\text{ mg/L}) = 51.7\text{ lb}$$

$$\textbf{Calcium hypochlorite, gal} = \frac{\mathbf{[Ca(OCl)_2,\ lb](100\%)}}{\mathbf{(10.21\ lb/gal)[Ca(OCl)_2,\ \%]}}$$

$$\text{Calcium hypochlorite, gal} = \frac{(51.7\text{ lb})(100\%)}{(10.21\text{ lb/gal})(65\%)}$$

$= 7.79$ gal, round to **7.8 gal of $Ca(OCl)_2$**

71. A small tank containing 500 gal of water needs to be disinfected in order to be put back in service. If the dosage needed is 50 mg/L, how many pounds of calcium hypochlorite (65.5% available chlorine) are required?

First, convert gallons to million gallons.

$$\text{Number of mil gal} = (500\text{ gal})(1\text{ M}/1{,}000{,}000) = 0.0005\text{ mil gal}$$

Then, determine the pounds of calcium hypochlorite needed.

The equation is:

$$\textbf{Number of lb, } \mathbf{Cl_2} = \frac{\mathbf{(mil\ gal)(Dosage,\ mg/L)(8.34\ lb/gal)}}{\mathbf{Percent\ purity}}$$

(continued)

$$\text{Number of lb, calcium hypochlorite} = \frac{(0.0005 \text{ mil gal})(50 \text{ mg/L})(8.34 \text{ lb/gal})}{65.5\%/100\%}$$

Number of lb, calcium hypochlorite = 0.318 lb,

round to **0.3 lb of calcium hypochlorite**

72. A 2.0-ft diameter pipe that is 1.35 mi long was disinfected with chlorine. If 64 lb of chlorine were used, what was the dosage in milligrams per liter?

First, find the number of feet in 1.4 mi.

$$\text{Number ft} = (5{,}280 \text{ ft/mi})(1.35 \text{ mi}) = 7{,}128 \text{ ft}$$

Next, find the volume in cubic feet for the pipe.

The equation is:

$$\textbf{Volume, gal} = \mathbf{(0.785)(Diameter)^2(Length, ft)(7.48 \ gal/ft^3)}$$

$$\text{Volume, ft}^3 = (0.785)(2.0 \text{ ft})(2.0 \text{ ft})(7{,}128 \text{ ft})(7.48 \text{ gal/ft}^3) = 167{,}417 \text{ gal}$$

Convert the number of gallons to million gallons.

$$\text{Million gallons} = \frac{167{,}417 \text{ gal}}{1{,}000{,}000/1 \text{ M}} = 0.167 \text{ mil gal}$$

Now, calculate the dosage.

The equation is:

$$\textbf{Dosage, mg/L} = \frac{\textbf{lb, chlorine}}{\textbf{(mil gal)(8.34 lb/gal)}}$$

$$\text{Dosage, mg/L} = \frac{64 \text{ lb}}{(0.167 \text{ mil gal})(8.34 \text{ lb/gal})}$$

= 45.95 mg/L, round to **46 mg/L of chlorine**

73. If 175 gal of a sodium hypochlorite solution with 7.5% available chlorine is used to treat 3.0 mil gal of water in a storage tank, determine the actual chlorine dosage in milligrams per liter.

First, calculate the number of chlorine pounds used.

The equation is:

$$\textbf{NaOCl, lb} = \frac{\textbf{(NaOCl, gal)(8.34 lb/gal)(NaOCl, \%)}}{\textbf{100\%}}$$

$$\text{NaOCl, lb} = \frac{(175 \text{ gal})(8.34 \text{ lb/gal})(7.5\%)}{100\%} = 109.46 \text{ lb}$$

Using the "pounds" equation, determine the actual sodium hypochlorite dosage.

Number of lb = (mil gal)(Dosage, mg/L)(8.34 lb/gal)

Rearrange the equation to solve for dosage.

$$\text{NaOCl dosage, mg/L} = \frac{\text{NaOCl lb}}{(\text{mil gal})(8.34 \text{ lb/gal})}$$

$$\text{NaOCl dosage, mg/L} = \frac{109.46 \text{ lb}}{(3.0 \text{ mil gal})(8.34 \text{ lb/gal})} = 4.37 \text{ mg/L, round to } \mathbf{4.4\ mg/L}$$

74. How many gallons of a 62.5% calcium hypochlorite solution was used to disinfect a storage tank that is 44.5 ft in diameter and 19.5 ft to the overflow (assuming tank was full of water) if the dosage was 49 mg/L? Assume the calcium hypochlorite solution weighs 10.05 lb/gal.

First, find the number of gallons of water in the tank.

The equation is:

Number of gal in tank = (0.785)(Diameter)^2(Depth or height, ft)(7.48 gal/ft^3)

Number of gal in tank = (0.785)(44.5 ft)(44.5 ft)(19.5 ft)(7.48 gal/ft^3) = 226,739 gal

Next, convert gallons to million gallons.

Number of mil gal = (226,739 gal)(1 M/1,000,000 gal) = 0.2267 mil gal

(continued)

Then, determine the number of pounds of calcium hypochlorite used.

The equation is:

$$\textbf{Number of lb} = \textbf{(mil gal)(Dosage, mg/L)(8.34 lb/gal)}$$

$$\text{Number of lb} = (0.2267 \text{ mil gal})(49 \text{ mg/L})(8.34 \text{ lb/gal})$$

$$= 92.64 \text{ lb of calcium hypochlorite}$$

Last, determine the number of gallons of calcium hypochlorite used.

$$\textbf{Calcium hypochlorite, gal} = \frac{\mathbf{[Ca(OCl)_2, lb](100\%)}}{\mathbf{(10.05\ lb/gal)[Ca(OCl)_2, \%]}}$$

$$\text{Calcium hypochlorite, gal} = \frac{(92.64 \text{ lb})(100\%)}{(10.05 \text{ lb/gal})(62.5\%)}$$

= 14.75 gal, round to **15 gal of $Ca(OCl)_2$**

75. A tank that is full of water and is 78.5 ft in diameter and 27.6 ft to the overflow is disinfected with a solution of calcium hypochlorite. If 201 lb of calcium hypochlorite were used, what was the dosage applied to the tank of water?

First, find the number of gallons of water in the tank.

The equation is:

$$\textbf{Number of gal in tank} = \mathbf{(0.785)(Diameter)^2(Depth\ or\ height,\ ft)(7.48\ gal/ft^3)}$$

$$\text{Number of gal in tank} = (0.785)(78.5 \text{ ft})(78.5 \text{ ft})(27.6 \text{ ft})(7.48 \text{ gal/ft}^3) = 998{,}665 \text{ gal}$$

Next, convert gallons to million gallons.

$$\text{Number of mil gal} = (998{,}665 \text{ gal})(1 \text{ M}/1{,}000{,}000 \text{ gal}) = 0.9987 \text{ mil gal}$$

Then, determine the dosage in milligrams per liter of calcium hypochlorite.

The equation is:

$$\textbf{Number of lb} = \textbf{(mil gal)(Dosage, mg/L)(8.34 lb/gal)}$$

Solve for dosage.

$$\text{Dosage, mg/L} = \frac{\text{Number of lb}}{\text{(mil gal)(8.34 lb/gal)}}$$

$$\text{Dosage, mg/L} = \frac{\text{201 lb}}{\text{(0.9987 mil gal)(8.34 lb/gal)}} = \text{24.13 mg/L, round to } \mathbf{24\ mg/L}$$

76. If 41 lb of calcium hypochlorite was found to be equivalent to 8.3 gal of hypochlorite solution and was used to treated 1.0 mil gal of water, what was the percent strength of the calcium hypochlorite solution?

$$\mathbf{Ca(OCl)_2\text{, gal} = \frac{[Ca(OCl)_2\text{, lb}][100\%\text{, }Ca(OCl)_2]}{[\text{Percent, }Ca(OCl)_2](8.34\text{ lb/gal})}}$$

Rearrange to solve for hypochlorite %.

$$Ca(OCl)_2\text{, \%} = \frac{[Ca(OCl)_2\text{, lb}][100\%\text{,}Ca(OCl)_2]}{[Ca(OCl)_2\text{, gal}](8.34\text{ lb/gal})}$$

$$Ca(OCl)_2\text{, \%} = \frac{\text{(41 lb)(100\%)}}{\text{(8.3 gal)(8.34 lb/gal)}}$$

= 59.2%, round to **59% calcium hypochlorite solution**

77. A 10.5% sodium hypochlorite solution is used to treat water from a well. The hypochlorinator is pumping at a rate of 11.5 gpd. What is the chlorine dosage in milligrams per liter if the well is producing 245 gpm?

First, convert the production rate of the pump to million gallons per day.

$$\mathbf{\text{Million gallons per day} = \frac{\text{(Pumping rate)(1,440 min/d)}}{\text{1,000,000/1 M}}}$$

$$\text{Million gallons per day} = \frac{\text{(245 gpm)(1,440 min/d)}}{\text{1,000,000/1 M}} = \text{0.3528 mgd}$$

(continued)

Next, calculate the sodium hypochlorite usage in pounds per day.

$$\textbf{NaOCl, lb/d} = \frac{\textbf{(NaOCl flow, gpd)(Percent, NaOCl)(8.34 lb/gal)}}{\textbf{100\%}}$$

$$\text{NaOCl, lb/d} = \frac{(11.5\text{ gpd})(10.5\%)(8.34\text{ lb/gal})}{100\%} = 10.07\text{ lb/d}$$

Last, calculate the sodium hypochlorite dosage in milligrams per liter using the "pounds" equation.

$$\text{Dosage, mg/L of NaOCl} = \frac{\text{NaOCl, lb/d}}{(\text{Well flow, mgd})(8.34\text{ lb/gal})}$$

$$\text{Dosage, mg/L of NaOCl} = \frac{10.07\text{ lb/d}}{(0.3528\text{ mgd})(8.34\text{ lb/gal})} = \mathbf{3.42\ mg/L\ of\ NaOCl}$$

78. A 750,000-gal water tank is to be disinfected with a 65% calcium hypochlorite solution. The dosage desired is 25 mg/L. How many liters of hypochlorite will be required? Assume the calcium hypochlorite solution weighs 10.21 lb/gal.

First, convert 750,000 gal to million gallons.

$$\frac{750{,}000\text{ gal}}{1{,}000{,}000/1\text{ M}} = 0.75\text{ mil gal}$$

Next, determine the number of pounds of hypochlorite required.

The equation is:

$$\textbf{Number of lb} = \frac{\textbf{(mil gal)(Dosage, mg/L)(8.34 lb/gal)}}{\textbf{Percent purity}}$$

Note: Delete "day" on each side of the equation as it is not needed.

$$Ca(OCl)_2\text{, lb} = \frac{(0.75\text{ mil gal})(25\text{ mg/L})(8.34\text{ lb/gal})}{65\%/100\%} = 240.58\text{ lb, } Ca(OCl)_2$$

Now, convert the number of pounds of calcium hypochlorite to gallons.

$$\frac{240.58\text{ lb, } Ca(OCl)_2}{10.21\text{ lb/gal}} = 23.56\text{ gal, } Ca(OCl)_2$$

Now, convert gallons to liters.

$$(23.56\text{ gal})(3.785\text{ L/gal}) = 89.17\text{ L of hypo, round to } \mathbf{89\ L\ of\ Ca(OCl)_2}$$

79. A pipe that is 3.0 ft in diameter and 872 ft long is to be disinfected with 6.0% sodium hypochlorite solution. If the desired dose is 25 mg/L, how many gallons of sodium hypochlorite are needed? The sodium hypochlorite solution weighs 9.02 lb/gal.

First, calculate the number of gallons in the pipeline.

The equation is:

$$\textbf{Number of gal} = \textbf{(0.785)(Diameter, ft)}^2\textbf{(Length, ft)(7.48 gal/ft}^3\textbf{)}$$

$$\text{Number of gal} = (0.785)(3.0\text{ ft})(3.0\text{ ft})(872\text{ ft})(7.48\text{ gal/ft}^3) = 46{,}082\text{ gal}$$

Convert gallons to million gallons.

$$\text{Number of mil gal} = (46{,}082\text{ gal})(1\text{ M}/1{,}000{,}000) = 0.0461\text{ mil gal}$$

Next, determine the number of pounds of sodium hypochlorite needed using the "pounds" equation.

$$\text{Number of lb, NaOCl} = (\text{mil gal})(\text{Dosage, mg/L})(8.34\text{ lb/gal})$$

$$\text{Number of lb, NaOCl} = (0.0461\text{ mil gal})(25\text{ mg/L})(8.34\text{ lb/gal}) = 9.61\text{ lb}$$

Last, determine the number of gallons needed.

$$\textbf{Number of gal, NaOCl} = \frac{\textbf{(Number of lb, NaOCl)(100\%)}}{\textbf{(9.02 lb/gal)(6.0\%)}}$$

$$\text{Number of gal, NaOCl} = \frac{(9.61\text{ NaOCl})(100\%)}{(9.02\text{ lb/gal})(6.0\%)} = 17.76\text{ gal, round to } \textbf{18 gal of NaOCl}$$

80. A pipe that is 2.5 ft in diameter and 2,950 ft long is to be disinfected with 64.0% calcium hypochlorite solution. If the desired dose is 50.0 mg/L, how many gallons of calcium hypochlorite are needed? Assume the hypochlorite solution weighs 10.15 lb/gal.

First, calculate the number of gallons in the pipeline.

The equation is:

$$\textbf{Number of gal} = \textbf{(0.785)(Diameter, ft)}^2\textbf{(Length, ft)(7.48 gal/ft}^3\textbf{)}$$

$$\text{Number of gal} = (0.785)(2.5\text{ ft})(2.5\text{ ft})(2{,}950\text{ ft})(7.48\text{ gal/ft}^3) = 108{,}261\text{ gal}$$

(continued)

Convert gallons to million gallons.

Number of mil gal = (108,261 gal)(1 M/1,000,000) = 0.1083 mil gal

Next, determine the number of pounds of calcium hypochlorite needed using the "pounds" equation.

$$\text{Number of lb, Ca(OCl)}_2 = \text{(mil gal)(Dosage, mg/L)(8.34 lb/gal)}$$

$$\text{Number of lb, Ca(OCl)}_2 = \text{(0.1083 mil gal)(50.0 mg/L)(8.34 lb/gal)} = 45.16 \text{ lb}$$

Last, determine the number of gallons needed.

$$\textbf{Number of gal, Ca(OCl)}_\mathbf{2} = \frac{\textbf{[Number of lb, Ca(OCl)}_\mathbf{2}\textbf{](100\%)}}{\textbf{(10.15 lb/gal)(Percent of solution)}}$$

$$\text{Number of gal, Ca(OCl)}_2 = \frac{[45.16 \text{ Ca(OCl)}_2](100\%)}{(10.15 \text{ lb/gal})(64.0\%)}$$

= 6.95 gal, round to **7.0 gal of Ca(OCl)$_2$**

81. An 18-in. diameter pipe that is 475 ft long is disinfected with a 60.5% calcium hypochlorite solution. If 0.52 gal were used, what must have been the dosage of the calcium hypochlorite? Because the hypochlorite has been diluted, assume it weighs 8.34 lb/gal.

First, convert the diameter of the pipe from inches to feet.

Number of ft = (18 in.)(1 ft/12 in.) = 1.5 ft

Next, determine the number of gallons in the pipeline.

Number of gal = (0.785)(1.5 ft)(1.5 ft)(475 ft)(7.48 gal/ft^3) = 6,275 gal

Then, convert gallons to million gallons.

Number of mil gal = (6,275 gal)(1 M/1,000,000) = 0.006275 mil gal

Next, determine the number of pounds used by rearranging the following equation to solve for pounds.

$$\textbf{Number of gal, Ca(OCl)}_\mathbf{2} = \frac{\textbf{[Number of lb, Ca(OCl)}_\mathbf{2}\textbf{](100\%)}}{\textbf{(8.34 lb/gal)(60.5\%)}}$$

$$\text{Number of lb, Ca(OCl)}_2 = \frac{[\text{Number of gal, Ca(OCl)}_2](8.34 \text{ lb/gal})(60.5\%)}{100\%}$$

$$\text{Number of lb, Ca(OCl)}_2 = \frac{[0.52 \text{ gal, Ca(OCl)}_2](8.34 \text{ lb/gal})(60.5\%)}{100\%} = 2.624 \text{ lb, Ca(OCl)}_2$$

Last, solve for the dosage of calcium hypochlorite using the "pounds" equation.

$$\text{Number of lb, Ca(OCl)}_2 = (\text{mil gal})(\text{Dosage, mg/L})(8.34 \text{ lb/gal})$$

$$\text{Dosage, mg/L of Ca(OCl)}_2 = \frac{\text{Number of lb, Ca(OCl)}_2}{(\text{mil gal})(8.34 \text{ lb/gal})}$$

$$\text{Dosage, mg/L of Ca(OCl)}_2 = \frac{2.624 \text{ lb, Ca(OCl)}_2}{(0.006275 \text{ mil gal})(8.34 \text{ lb/gal})}$$

Dosage, mg/L of $Ca(OCl)_2$ = 50.1 mg/L, round to **50 mg/L of $Ca(OCl)_2$**

82. A trough that is 50 ft in length, has a bottom width of 5 ft, a top of water width of 9 ft, and a depth of water of 4 ft is disinfected with a 5.0% sodium hypochlorite solution. How many gallons of sodium hypochlorite will be required if a 30.0-mg/L dosage is desired?

First, find the volume in water in the trough.

The equation is:

$$\textbf{Volume, gal} = \frac{(b_1 + b_2)}{2}\textbf{(Depth of water, ft)(Length, ft)(7.48 gal/ft}^3\textbf{)}$$

$$\text{Volume, ft}^3 = \frac{(5 \text{ ft} + 9 \text{ ft})}{2}(4 \text{ ft})(50 \text{ ft})(7.48 \text{ gal/ft}^3) =$$

$$\text{Volume, ft}^3 = \frac{(14 \text{ ft})(4 \text{ ft})(50 \text{ ft})(7.48 \text{ gal/ft}^3)}{2} = \frac{20{,}944 \text{ gal}}{2} = 10{,}472 \text{ gal}$$

Next, convert gallons to million gallons.

$$\text{Number of mil gal} = (10{,}472 \text{ gal})(1 \text{ M}/1{,}000{,}000) = 0.01047 \text{ mil gal}$$

Next, find the number of pounds of sodium hypochlorite needed.

$$\text{Number of lb, NaOCl} = (\text{mil gal})(\text{Dosage, mg/L})(8.34 \text{ lb/gal})$$

$$\text{Number of lb, NaOCl} = (0.01047 \text{ mil gal})(30 \text{ mg/L})(8.34 \text{ lb/gal}) = 2.62 \text{ lb}$$

(continued)

Last, determine the number of gallons needed.

$$\text{Number of gal, NaOCl} = \frac{\text{(Number of lb, NaOCl)(100\%)}}{\text{(8.34 lb/gal)(5.0\%)}}$$

$$\text{Number of gal, NaOCl} = \frac{\text{(2.62 NaOCl)(100\%)}}{\text{(8.34 lb/gal)(5.0\%)}} = \text{6.28 gal, round to } \textbf{6 gal of NaOCl}$$

83. A 24-in. diameter pipe that is 1,618 ft long is disinfected with a 62.5% calcium hypochlorite solution. If 3.03 gal were used, what must have been the dosage of the calcium hypochlorite? Assume the calcium hypochlorite weighs 8.34 lb/gal.

First, convert the diameter of the pipe from inches to feet.

$$\text{Number of ft} = \text{(24 in.)(1 ft/12 in.)} = \text{2.0 ft}$$

Next, determine the number of gallons in the pipeline.

$$\text{Number of gal} = \text{(0.785)(2.0 ft)(2.0 ft)(1,618 ft)(7.48 gal/ft}^3\text{)} = \text{38,002 gal}$$

Then, convert gal to million gallons.

$$\text{Number of mil gal} = \text{(38,002 gal)(1 M/1,000,000)} = \text{0.038 mil gal}$$

Next, determine the number of pounds used by rearranging the following equation to solve for pounds.

$$\textbf{Number of gal, Ca(OCl)}_2 = \frac{\textbf{[Number of lb, Ca(OCl)}_2\textbf{](100\%)}}{\textbf{(8.34 lb/gal)(62.5\%)}}$$

$$\text{Number of lb, Ca(OCl)}_2 = \frac{\text{[Number of gal, Ca(OCl)}_2\text{](8.34 lb/gal)(62.5\%)}}{100\%}$$

$$\text{Number of lb, Ca(OCl)}_2 = \frac{\text{[3.03 gal, Ca(OCl)}_2\text{](8.34 lb/gal)(62.5\%)}}{100\%} = \text{15.79 lb, Ca(OCl)}_2$$

Last, solve for the dosage of calcium hypochlorite using the "pounds" equation.

$$\text{Number of lb, Ca(OCl)}_2 = \text{(mil gal)(Dosage, mg/L)(8.34 lb/gal)}$$

$$\text{Dosage, mg/L of Ca(OCl)}_2 = \frac{\text{Number of lb, Ca(OCl)}_2}{\text{(mil gal)(8.34 lb/gal)}}$$

$$\text{Dosage, mg/L of Ca(OCl)}_2 = \frac{\text{15.79 lb, Ca(OCl)}_2}{\text{(0.038 mil gal)(8.34 lb/gal)}}$$

Dosage, mg/L of $Ca(OCl)_2$ = 49.82 mg/L, round to **50 mg/L of $Ca(OCl)_2$**

PERCENT MIXTURE PROBLEMS

84. If 225 lb of high test hypochlorite (HTH) with 63.5% available chlorine are mixed with 250 gal of water, what percent is the calcium hypochlorite solution? What percent solution of 100% available chlorine is this?

The equation is:

$$\textbf{Percent HTH solution} = \frac{\textbf{(lb HTH)(100\%)}}{\textbf{(Number of gal)(8.34 lb/gal)}}$$

$$\text{Percent HTH solution} = \frac{\text{(225 lb HTH)(100\%)}}{\text{(250 gal)(8.34 lb/gal)}} = 10.79\%, \text{ round to } \textbf{10.8\% HTH solution}$$

Percent available chlorine solution = (10.79%)(63.5%/100%)

Percent available chlorine solution = 6.85 %, round to **6.8 % available chlorine**

85. How many pounds of high test hypochlorite are needed to make exactly 150 gal of a 5.0% high test hypochlorite solution?

The equation is:

$$\textbf{Percent HTH solution} = \frac{\textbf{(lb HTH)(100\%)}}{\textbf{(Number of gal)(8.34 lb/gal)}}$$

Rearrange the equation.

Number of lb, HTH = (Percent solution)(Number of gal)(8.34 lb/gal)/100%

Number of lb, HTH = (5.0% solution)(150 gal)(8.34 lb/gal)/100% =

Number of lb, HTH = 62.55 lb, round to **63 lb, HTH**

Dilution Triangle

The dilution triangle is probably the easiest way to solve dilution problems. The following two problems demonstrate its usage.

To solve the problems using the dilution triangle follow the setups as shown and the following instructions: The two numbers on the left are the existing concentrations of 15% and 2.1%. The number in the center, 8%, is the desired concentration. The numbers on the right are determined by subtracting diagonally the existing concentrations from the desired concentration.

86. How many gallons of a 15.0% solution must be mixed with a 2.1% solution to make exactly 2,500 gal of an 8.0% solution?

Solve the problem using the dilution triangle.

15.0% 5.9* 5.9 parts of the 15.0% solution are required for every 12.9 parts

8.0%

2.1% 7.0† 7.0 parts of the 2.1% solution are required for every 12.9 parts

12.9 total parts

$$\frac{5.9 \text{ parts}(2{,}500 \text{ gal})}{12.9 \text{ parts}} = \textbf{1,143 gal of the 15.0\% solution}$$

$$\frac{7.0 \text{ parts } (2{,}500 \text{ gal})}{12.9 \text{ parts}} = \textbf{1,357 gal of the 2.1\% solution}$$

2,500 gal (added as a cross-check of math)

To make the 2,500 gal of the 8.0% solution, mix 1,143 gal of the 15.0% solution with 1,357 gal of the 2.1% solution.

* This number is calculated by subtracting diagonally 8.0% from 2.1%, absolute (negative sign is dropped).

† This number is calculated by subtracting diagonally 8.0% from 15.0%.

87. How many gallons of a 15% solution must be mixed with a 2.5% solution to make exactly 1,000 gal of a 4.0% solution?

Solve the problem using the dilution triangle.

15% 1.5 1.5 parts of the 15% solution are required for every 12.5 parts

4.0%

2.5% 11 11 parts of the 2.5% solution are required for every 12.5 parts

12.5 total parts

$$\frac{1.5 \text{ parts}(1{,}000 \text{ gal})}{12.5 \text{ parts}} = \textbf{120 gal of the 15\% solution}$$

$$\frac{11 \text{ parts } (1{,}000 \text{ gal})}{12.5 \text{ parts}} = \frac{\textbf{880}}{1{,}000 \text{ gal}} \textbf{ gal of the 2.5\% solution}$$

To make the 1,000 gal of the 4.0% solution, mix 120 gal of the 15% solution with 880 gal of the 2.5% solution.

SPECIFIC YIELD AND DRAWDOWN CALCULATIONS

88. A well produces 136 gpm. If the drawdown for the well is 12.3 ft, find the specific yield in gallons per minute per foot.

The equation is:

$$\textbf{Specific yield, gpm/ft} = \frac{\textbf{Well yield, gpm}}{\textbf{Drawdown, ft}}$$

$$\text{Specific yield, gpm/ft} = \frac{136 \text{ gpm}}{12.3 \text{ ft}} = 11.057 \text{ gpm/ft, round to } \textbf{11 gpm/ft}$$

89. Find the drawdown of a well if the well yields 188 gpm and the specific yield is 12.7.

Write the equation, arranging it to solve for the unknown, drawdown.

The equation is:

$$\textbf{Drawdown, ft} = \frac{\textbf{Well yield, gpm}}{\textbf{Specific yield, gpm/ft}}$$

$$\text{Drawdown, ft} = \frac{188 \text{ gpm}}{12.7 \text{ gpm/ft}} = \textbf{14.8 ft}$$

90. Determine the specific yield in gallons per minute per foot for a well that produces an average of 218 gpm and has a drawdown of 21.5 ft.

The equation is:

$$\textbf{Specific yield, gpm/ft} = \frac{\textbf{Well yield, gpm}}{\textbf{Drawdown, ft}}$$

$$\text{Specific yield, gpm/ft} = \frac{218 \text{ gpm}}{21.5 \text{ ft}} = 10.14 \text{ gpm/ft, round to } \textbf{10.1 gpm/ft}$$

WELL DRAWDOWN CALCULATIONS

91. If the static water level (nonpumping well) in a well is 102.8 ft, what is the drawdown if the pumping level is 116.9 ft?

The equation is:

$$\textbf{Drawdown, ft} = \textbf{Pumping water level, ft} - \textbf{Static water level, ft}$$

$$\text{Drawdown, ft} = 116.9 \text{ ft} - 102.8 \text{ ft} = \textbf{14.1 ft}$$

92. If the static level in the well was 99.35 ft and the drawdown was 15.56 ft, what must have been the pumping water level in the well?

The equation is:

$$\text{Drawdown, ft} = \text{Pumping water level, ft} - \text{Static water level, ft}$$

Rearrange the equation to solve for pumping water level.

$$\textbf{Pumping water level, ft} = \textbf{Drawdown, ft} + \textbf{Static water level}$$

Substitute known values.

$$\text{Pumping water level, ft} = 99.35 \text{ ft} + 15.56 \text{ ft} = 114.91 \text{ ft, round to } \mathbf{114.9 \text{ ft}}$$

93. The pumping water level in the well is 95.26 ft. If the drawdown was 17.11 ft, what must have been the static water level in the well?

The equation is:

$$\textbf{Drawdown, ft} = \textbf{Pumping water level, ft} - \textbf{Static water level, ft}$$

Rearrange the equation to solve for static water level.

$$\text{Static water level, ft} = \text{Pumping water level, ft} - \text{Drawdown, ft}$$

Substitute known values.

$$\text{Static water level, ft} = 95.26 \text{ ft} - 17.11 \text{ ft} = \mathbf{78.15 \text{ ft}}$$

PUMP DISCHARGE PROBLEMS

94. If a pump discharges 7,880 gal in 2 hr and 13 min, how many gallons per minute is the pump discharging?

First, determine the number of minutes the pump was working.

$$2 \text{ hr}(60 \text{ min/hr}) + 13 \text{ min} = 120 \text{ min} + 13 \text{ min} = 133 \text{ min}$$

Then, determine the number of gallons per minute by dividing the number of gallons pumped by the total time the pump worked.

$$\frac{7{,}880 \text{ gal}}{133 \text{ min}} = 59.25 \text{ gpm, round to } \mathbf{59 \text{ gpm}}$$

95. If a pump discharges 375 gpm, how many gallons will it discharge in 12 hr?

First, determine the number of minutes the pump was working.

$$12 \text{ hr}(60 \text{ min/hr}) = 720 \text{ min}$$

Then, determine the number of gallons that it will discharge in 12 hr.

$$\textbf{Number of gal discharged} = \textbf{(Number of min)(Pump rate, gpm)}$$

$$\text{Number of gal discharged} = (720 \text{ min})(375 \text{ gpm}) = \textbf{270,000 gal}$$

96. A hypochlorite solution is being pumped from a small tank that is 2.5 ft in diameter. If the level in the tank drops 2.05 ft in 3.5 hr, how many gallons per minute of hypochlorite solution was used?

First, determine the number of gallons that was pumped.

The equation is:

$$\textbf{Number of gal} = \textbf{(0.785)(Diameter, ft)}^2\textbf{(Level drop, ft)(7.48 gal/ft}^3\textbf{)}$$

$$\text{Number of gal} = (0.785)(2.5 \text{ ft})(2.5 \text{ ft})(2.05 \text{ ft})(7.48 \text{ gal/ft}^3) = 75.23 \text{ gal}$$

Next, find the number of minutes in 3.5 hr.

$$\text{Number of min} = (3.5 \text{ hr})(60 \text{ min/hr}) = 210 \text{ min}$$

Last, divide the number of gallons pumped by the time in minutes.

$$\text{Number of gpm} = \frac{75.23 \text{ gal}}{210 \text{ min}} = \textbf{0.36 gpm}$$

PUMPING CALCULATIONS

97. Find the total head in feet for a pump with a total static head of 16 ft and a head loss of 4.6 ft.

The equation is:

$$\textbf{Total head, ft} = \textbf{Total static head, ft} + \textbf{Head losses, ft}$$

$$\text{Total head, ft} = 16 \text{ ft} + 4.6 \text{ ft} = 20.6 \text{ ft, round to } \textbf{21 ft}$$

98. Water is being pumped from a water source that has an elevation of 4,438 ft to an elevation of 4,922 ft. What is the total head if friction and minor head losses are 23.0 ft?

The equation is:

Total head, ft = Total static head, ft (difference in elevation) + Head losses, ft

Total head, ft = (4,922 ft – 4,438 ft) + 23.0 ft = 484 + 23.0 ft = **507 ft**

99. Find the motor horsepower for a pump station with the following parameters:

Motor efficiency:	**87%**	**Total head:**	**145 ft**
Pump efficiency:	**79%**	**Flow:**	**1.68 mgd**

First, convert million gallons per day to gallons per minute.

Gallons per minute = (1.68 mgd)(1,000,000/1 M)(1 d/1,440 min) = 1,167 gpm

The equation is:

$$\textbf{Motor horsepower} = \frac{\textbf{(Flow, gpm)(Total head, ft)}}{\textbf{(3,960)(Motor efficiency)(Pump efficiency)}}$$

Note: The 3,960 is a constant.

Next, convert the percent motor and pump efficiencies to decimal form by dividing by 100%.

Motor efficiency = 87%/100% = 0.87

Pump efficiency = 79%/100% = 0.79

$$\text{Motor horsepower} = \frac{(1{,}167 \text{ gpm})(145 \text{ ft})}{(3{,}960)(0.87 \text{ Motor efficiency})(0.79 \text{ Pump efficiency})} = \textbf{62 mhp}$$

100. What is the motor horsepower if 100 hp is required to run a pump with a motor efficiency of 88% and a pump efficiency of 78%?

Note: The 100 hp in this problem is called the water horsepower (whp). The whp is the actual energy (horsepower) available to pump water. Assume 100 hp has two significant figures.

The equation is:

$$\textbf{Motor horsepower} = \frac{\textbf{(whp)}}{\textbf{(Motor efficiency)(Pump efficiency)}}$$

First, change percent motor and pump efficiencies to decimal form by dividing by 100%.

Motor efficiency = 88%/100% = 0.88

Pump efficiency = 78%/100% = 0.78

Or, it can be done within the problem as follows:

$$\text{Motor horsepower} = \frac{\text{(100 whp)}}{\text{(88\%/100\% Motor efficiency)(78\%/100\% Pump efficiency)}}$$

$$\text{Motor horsepower} = \frac{\text{(100 whp)}}{\text{(0.88 Motor efficiency)(0.78 Pump efficiency)}}$$

Motor horsepower = 145.69 mhp, round to **150 mhp**

101. What is the brake horsepower if 62 hp is supplied to a motor with 87% efficiency?

The equation is:

Brake horsepower = (hp)(Motor efficiency)

Brake horsepower = (62 hp)(87%/100% Motor efficiency) = 53.94 bhp, round to **54 bhp**

102. Find the water horsepower for the following system:

Motor efficiency is 90%.

Pump efficiency is 79%.

Motor horsepower is 80.

The equation is:

$$\textbf{Water horsepower} = \textbf{(mhp)(Motor efficiency)(Pump efficiency)}$$

First, convert the motor and pump efficiencies to decimal form.

$$\text{Motor efficiency} = 90\%/100\% = 0.90$$

$$\text{Pump efficiency} = 79\%/100\% = 0.79$$

$$\begin{aligned}\text{Water horsepower} &= (80 \text{ mhp})(0.90 \text{ Motor efficiency})(0.79 \text{ Pump efficiency}) \\ &= 56.88 \text{ whp, round to } \mathbf{57\ whp}\end{aligned}$$

103. What is the motor horsepower if the brake horsepower is 68 and the motor efficiency is 87%?

The equation is:

$$\textbf{Motor horsepower} = \frac{\textbf{bhp}}{\textbf{Motor efficiency}}$$

$$\text{Motor efficiency} = 87\%/100\% = 0.87$$

$$\text{Motor horsepower} = \frac{68 \text{ bhp}}{0.87 \text{ Motor efficiency}} = 78.16 \text{ mhp, round to } \mathbf{78\ mhp}$$

104. Given the following parameters, determine the motor horsepower for a pump station:

Motor efficiency:	**88.0%**	**Total head:**	**108 ft**
Pump efficiency:	**75.5%**	**Flow:**	**2.27 mgd**

First, convert million gallons per day to gallons per minute.

Gallons per minute = (2.27 mgd)(1,000,000/1 M)(1 d/1,440 min) = 1,576 gpm

Next, convert the percent motor and pump efficiencies to decimal form by dividing by 100%.

Motor efficiency = 88.0%/100% = 0.880

Pump efficiency = 75.5%/100% = 0.755

The equation for determining the motor horsepower with the given data is different then the problem above.

The equation is:

$$\textbf{Motor horsepower} = \frac{\textbf{(Flow, gpm)(Total head, ft)}}{\textbf{(3,960)(Motor efficiency)(Pump efficiency)}}$$

Note: The number 3,960 is a constant.

The equation is:

$$\text{Motor horsepower} = \frac{(1{,}576 \text{ gpm})(108 \text{ ft})}{(3{,}960)(0.880 \text{ Motor efficiency})(0.755 \text{ Pump efficiency})}$$

$$= \mathbf{64.7\ mhp}$$

105. Calculate the water horsepower for the following system:

Motor efficiency is 90.8%.

Pump efficiency is 76.5%.

Motor horsepower is 148.

The equation is:

Water horsepower = (mhp)(Motor efficiency)(Pump efficiency)

Motor efficiency = 90.8%/100% = 0.908

Pump efficiency = 76.5%/100% = 0.765

Water horsepower = (148 mhp)(0.908 Motor efficiency)(0.765 Pump efficiency)

Water horsepower = 102.8 whp, round to **103 whp**

WATER RATE SERVICE PROBLEMS

106. What is the quarterly billing rate for a family of four given the following data? Assume 30 days per month.

Usage in gallons per person:	**120 gal**
Basic service rate:	**\$8.00/quarter**
The first 1,000 ft^3 used is:	**\$0.40/ccf**
Additional 1,000 ft^3 used is:	**\$0.65/ccf**

First, determine the number of cubic feet the family uses, where month = mo.

$$\text{Number ft}^3 = \frac{(4\text{ people/quarter})(120\text{ gal/person/d})(30\text{ d/mo})(3\text{ mo/quarter})}{7.48\text{ gal/ft}^3} = 5{,}775\text{ ft}^3$$

$$\text{The first }1{,}000\text{ ft}^3\text{ rate} = \frac{(1{,}000\text{ ft}^3)(\$0.40)}{100\text{ ccf}} = \$4.00$$

Next, determine the additional pay rate.

$$\text{Total used, ft}^3 - \text{First, }1{,}000\text{ ft}^3 = \text{Additional rate usage, ft}^3$$

$$5{,}775\text{ ft}^3 - 1{,}000\text{ ft}^3 = 4{,}775\text{ ft}^3$$

$$\text{The first }1{,}000\text{ ft}^3\text{ rate} = \frac{(4{,}775\text{ ft}^3)(\$0.65)}{100\text{ ccf}} = \$31.04$$

$$\text{Total bill} = \$8.00 + \$4.00 + \$31.04 = \mathbf{\$43.04}$$

107. What is the quarterly billing rate for a farm given the following data? Assume 30 days per month.

Average usage in gallons per day:	**10,000 gal**
Basic service rate:	**\$10.00/quarter**
The first 100,000 ft^3 used is:	**\$0.05/ccf**
Additional 100,000 ft^3 used is:	**\$0.10/ccf**

First, determine the number of cubic feet the farm uses.

$$\text{Number ft}^3 = \frac{(10{,}000\text{ gpd/quarter})(30\text{ d/mo})(3\text{ mo/quarter})}{7.48\text{ gal/ft}^3} = 120{,}321\text{ ft}^3$$

$$\text{The first }100{,}000\text{ ft}^3\text{ rate} = \frac{(100{,}000\text{ ft}^3)(\$0.05)}{100\text{ ccf}} = \$50.00$$

Next, determine the additional pay rate.

$$\text{Total used, ft}^3 - \text{First, 1000 ft}^3 = \text{Additional rate usage, ft}^3$$

$$120{,}321 \text{ ft}^3 - 100{,}000 \text{ ft}^3 = 20{,}321 \text{ ft}^3$$

$$\text{The first 1,000 ft}^3 \text{ rate} = \frac{(20{,}321 \text{ ft}^3)(\$0.10)}{100 \text{ ccf}} = \$20.32$$

$$\text{Total bill} = \$10.00 + \$50.00 + \$20.32 = \mathbf{\$82.32}$$

Distribution Grade 3 Test

(answers on p. 162)

1. If 50.0 lb of soda ash are mixed into 248 gal of water, what is the percent of soda ash in the slurry?

2. A distribution pipe is 4.57 mi long. What is the volume of water in gallons if the pipe is 3.0 ft in diameter for a length of 0.95 mi and 18 in. for the remainder?

3. Calculate the detention time in hours for the following water treatment plant's clear well and associated distribution line and storage tank:

   The clear well is 498.5 ft long, 210.3 ft wide, and has an average water depth of 11.5 ft. The distribution line from clear well to storage tank is 1.01 mi long and has a diameter of 2.00 ft. The storage tank has a diameter of 75.0 ft and averages a height of 29.8 ft of water. The flow is 25.2 mgd for the day in question.

4. A 10.0-in. diameter distribution pipe delivers 1,294,000 gal in exactly 24 hr. What is the average flow rate during the 24-hr time period in feet per second?

5. A 27.8-ft, ¾ in. service line requires flushing. How many minutes are required to flush the line 20 volumes if the line is flushed at a rate of 12.0 gpm?

6. A 3.00-ft diameter pipe that is 1.26 mi long was disinfected with chlorine. If 73.2 lb of 100% available chlorine were used, what was the dosage in milligrams per liter?

7. How many gallons of a 64.25% calcium hypochlorite solution was used to disinfect a storage tank that is 77.5 ft in diameter and 29.5 ft to the overflow (assuming tank was full of water) if the dosage was 50.6 mg/L? Assume the calcium hypochlorite solution weighs 10.17 lb/gal.

8. An 8.30% sodium hypochlorite solution is treating water from a well. The hypochlorinator is pumping at a rate of 7.58 gpd. What is the chlorine dosage in milligrams per liter if the well is producing 305 gpm? Assume the sodium hypochlorite solution weighs 9.30 lb/gal.

9. A pipe that is 3.00 ft in diameter and 956 ft long is to be disinfected with 60.0% calcium hypochlorite solution. If the desired dose is 40.0 mg/L, how many gallons of calcium hypochlorite are needed? Assume the calcium hypochlorite solution weighs 9.98 lb/gal.

10. How many gallons of a 13.0% solution must be mixed with a 3.50% solution to make exactly 2,500 gal of a 7.50% solution?

Solutions to Distribution Grade 3 Test

1. First, find the number of pounds in 248 gal of water.

$$\text{Number of lb} = (8.34\ \text{lb/gal})(248\ \text{gal}) = 2{,}068.32\ \text{lb}$$

Next, determine the soda ash slurry.

$$\text{Percent soda ash} = \frac{(50.0\ \text{lb})(100\%)}{50.0\text{lb} + 2{,}068.32\ \text{lb}} = \frac{(50.0\ \text{lb})(100\%)}{2{,}118.32\ \text{lb}} = \mathbf{2.36\%\ soda\ ash}$$

2. The equation is:

Volume, gal = (Cross-sectional area of the pipe, ft^2)(Length, ft)(7.48 gal/ft^3)

Therefore, we must first convert miles and inches to feet in order to use the above equation.

First, convert 0.95 miles to ft for the larger pipe.

$$(5{,}280\ \text{ft/mi})(0.95\ \text{mi}) = 5{,}016\ \text{ft}$$

Then, determine the remainder of the feet for the smaller pipe.

Thus:

$$4.57\ \text{mi} - 0.95\ \text{mi} = 3.62\ \text{mi for smaller pipe.}$$

Convert the 3.62 mi to feet.

$$(5{,}280\ \text{ft/mi})(3.62\ \text{mi}) = 19{,}114\ \text{ft}$$

Next, convert 18 in. to feet.

$$(18.0\ \text{in.})(1\ \text{ft}/12\ \text{in.}) = 1.50\ \text{ft}$$

Then, determine the cross-sectional area of each pipe size.

The equation is:

$$\textbf{Area, ft}^2 = \textbf{(0.785)(Diameter)}^2$$

Cross-sectional area for 3.0-ft diameter pipe, ft^2 = (0.785)(3.0 ft)(3.0 ft) = 7.065 ft^2

Cross-sectional area for 1.5-ft diameter pipe, ft^2 = (0.785)(1.50 ft)(1.50 ft) = 1.766 ft^2

Volume, gal for 3.0-ft diameter pipe = (7.065 ft^2)(5,016 ft)(7.48 gal/ft^3) = 265,077 gal

Volume, gal for 1.5-ft diameter pipe = (1.766 ft^2)(19,114 ft)(7.48 gal/ft^3) = 252,490 gal

Last, add the two volumes together for the total volume.

Total volume, gal = 265,077 gal + 252,490 gal = 517,567 gal, round to **520,000 gal**

3. First, determine the number of gallons in each of the containment structures.

The equation is:

Volume, gal = (Length)(Width)(Depth)(7.48 gal/ft^3)

Volume, gal, clear well = (498.5 ft)(210.3 ft)(11.5 ft)(7.48 gal/ft^3) = 9,017,868 gal

Next, convert number of miles to feet.

Number of ft = (5,280 ft/mi)(1.01 mi) = 5,333 ft

Use the following equation:

Volume, gal = (Length)(0.785)(Diameter)^2(7.48 gal/ft^3)

Volume, gal, pipeline = (5,333 ft)(0.785 ft)(2.00 ft)(2.00 ft)(7.48 gal/ft^3) = 125,257 gal

Volume, gal, in filters = (0.785)(75.0 ft)(75.0 ft)(29.8 ft)(7.48 gal/ft^3) = 984,260 gal

10,127,385 gal

Next, convert million gallons per day to gallons per hour.

(25.2 mgd)(1 d/24 hr)(1,000,000/1 M) = 1,050,000 gph

Write the equation with units asked for in the question.

$$\textbf{Detention time, hr} = \frac{\textbf{Volume, gal}}{\textbf{Flow rate, gph}}$$

$$\text{Detention time, hr} = \frac{10{,}127{,}385 \text{ gal}}{1{,}050{,}000 \text{ gph}} = \mathbf{9.65\ hr}$$

4. First, convert the diameter in inches to feet.

$$\text{Number of ft} = (10.0 \text{ in.})(1 \text{ ft}/12 \text{ in.}) = 0.833$$

Next, determine the cross-sectional area of the pipe.

The equation is:

$$\mathbf{Area,\ ft^2 = (0.785)(Diameter)^2}$$

$$\text{Area, ft}^2 = (0.785)(0.833 \text{ ft})(0.833 \text{ ft}) = 0.545 \text{ ft}^2$$

Next, determine the cubic feet per second.

$$\text{Number of cubic feet per second} = (1{,}294{,}000 \text{ gal}/24 \text{ hr})(1 \text{ ft}^3/7.48 \text{ gal})(1 \text{ hr}/3{,}600 \text{ s})$$

$$= 2.00 \text{ ft}^3/\text{s}$$

Last, determine the flow in feet per second.

The equation is:

$$\mathbf{Flow,\ ft^3/s = (Area,\ ft^2)(Velocity,\ ft/s)}$$

Rearrange the formula to solve for velocity.

$$\text{Velocity, ft/s} = \frac{\text{Flow, ft}^3/\text{s}}{\text{Area, ft}^2}$$

Substituting:

$$\text{Velocity, ft/s} = \frac{2.00 \text{ ft}^3/\text{s}}{0.545 \text{ ft}^2} = \mathbf{3.67\ ft/s}$$

5. First, convert the pipe diameter from inches to feet.

$$\text{Number of ft} = (\tfrac{3}{4} \text{ in.})(1 \text{ ft}/12 \text{ in.}) = (0.75 \text{ in.})(1 \text{ ft}/12 \text{ in.}) = 0.0625 \text{ ft}$$

Next, determine the volume in gallons for 27.8 ft of the ¾-in. pipeline.

The equation is:

$$\mathbf{Volume,\ gal = (0.785)(Diameter)^2(Length)(7.48\ gal/ft^3)(25\ volumes)}$$

$$\text{Volume, gal} = (0.785)(0.0625 \text{ ft})(0.0625 \text{ ft})(27.8 \text{ ft})(7.48 \text{ gal/ft}^3)(20 \text{ volumes}) = 12.75 \text{ gal}$$

Then, find the flushing time.

$$\text{Flushing time, min} = \frac{12.75 \text{ gal}}{12.0 \text{ gpm}} = 1.06 \text{ min}$$

6. First, find the number of feet in 1.26 mi.

Number of ft = (5,280 ft/mi)(1.26 mi) = 6,653 ft

Next, find the volume in cubic feet for the pipe.

The equation is:

$$\textbf{Volume, gal} = \textbf{(0.785)(Diameter)}^2\textbf{(Length, ft)(7.48 gal/ft}^3\textbf{)}$$

$$\text{Volume, ft}^3 = (0.785)(3.00\text{ ft})(3.00\text{ ft})(6{,}653\text{ ft})(7.48\text{ gal/ft}^3) = 351{,}586\text{ gal}$$

Convert the number of gallons to million gallons.

$$\text{Million gallons} = \frac{351{,}586\text{ gal}}{1{,}000{,}000/1\text{ M}} = 0.3516\text{ mil gal}$$

Last, calculate the dosage.

$$\textbf{Dosage, mg/L} = \frac{\textbf{lb of chlorine}}{\textbf{(mil gal)(8.34 lb/gal)}}$$

$$\text{Dosage, mg/L} = \frac{73.2\text{ lb}}{(0.3516\text{ mil gal})(8.34\text{ lb/gal})}$$

= 24.96 mg/L, round to **25 mg/L, chlorine**

7. First, find the number of gallons of water in the tank.

The equation is:

$$\textbf{Number of gal in tank} = \textbf{(0.785)(Diameter)}^2\textbf{(Depth or height, ft)(7.48 gal/ft}^3\textbf{)}$$

$$\text{Number of gal in tank} = (0.785)(77.5\text{ ft})(77.5\text{ ft})(29.5\text{ ft})(7.48\text{ gal/ft}^3) = 1{,}040{,}391\text{ gal}$$

Next, convert gallons to million gallons.

$$\text{Number of mil gal} = (1{,}040{,}391\text{ gal})(1\text{ M}/1{,}000{,}000\text{ gal}) = 1.0404\text{ mil gal}$$

Then, determine the number of pounds of calcium hypochlorite used.

The equation is:

$$\textbf{Number of lb} = \textbf{(mil gal)(Dosage, mg/L)(8.34 lb/gal)}$$

$$\text{Number of lb} = (1.0404\text{ mil gal})(50.6\text{ mg/L})(8.34\text{ lb/gal}) = 439\text{ lb of calcium hypochlorite}$$

(continued)

Last, determine the number of gallons of calcium hypochlorite used.

$$\textbf{Calcium hypochlorite, gal} = \frac{\mathbf{[Ca(OCl)_2,\ lb](100\%)}}{\mathbf{(10.17\ lb/gal)[Ca(OCl)_2,\ \%]}}$$

$$\text{Calcium hypochlorite, gal} = \frac{(439\text{ lb})(100\%)}{(10.17\text{ lb/gal})(64.25\%)}$$

$$= 67.18\text{ gal, round to } \mathbf{67.2\ gal\ of\ Ca(OCl)_2}$$

8. First, convert the production rate of the pump to million gallons per day.

$$\textbf{Million gallons per day} = \frac{\textbf{(Pumping rate)(1,440 min/d)}}{\textbf{1,000,000/1 M}}$$

$$\text{Million gallons per day} = \frac{(305\text{ gpm})(1{,}440\text{ min/d})}{1{,}000{,}000/1\text{ M}} = 0.4392\text{ mgd}$$

Next, calculate the sodium hypochlorite usage in pounds per day.

$$\textbf{NaOCl, lb/d} = \frac{\textbf{(NaOCl flow, gpd)(Percent, NaOCl)(9.30 lb/gal)}}{\mathbf{100\%}}$$

$$\text{NaOCl, lb/d} = \frac{(7.58\text{ gpd})(8.30\%)(9.30\text{ lb/gal})}{100\%} = 5.85\text{ lb/d, NaOCl}$$

Last, calculate the sodium hypochlorite dosage in milligrams per liter using the "pounds" equation.

$$\text{Dosage, mg/L of NaOCl} = \frac{\text{NaOCl, lb/d}}{(\text{Well flow, mgd})(8.34\text{ lb/gal})}$$

$$\text{Dosage, mg/L of NaOCl} = \frac{5.85\text{ lb/d}}{(0.4392\text{ mgd})(8.34\text{ lb/gal})} = \textbf{1.60 mg/L of NaOCl}$$

9. First, calculate the number of gallons in the pipeline.

The equation is:

$$\textbf{Number of gal} = \textbf{(0.785)(Diameter, ft)}^2\textbf{(Length, ft)(7.48 gal/ft}^3\textbf{)}$$

$$\text{Number of gal} = (0.785)(3.00\text{ ft})(3.00\text{ ft})(956\text{ ft})(7.48\text{ gal/ft}^3) = 50{,}521\text{ gal}$$

Convert gallons to million gallons.

$$\text{Number of mil gal} = (50{,}521\text{ gal})(1\text{ M}/1{,}000{,}000) = 0.0505\text{ mil gal}$$

Next, determine the number of pounds of calcium hypochlorite needed using the "pounds" equation.

$$\text{Number of lb, Ca(OCl)}_2 = \text{(mil gal)(Dosage, mg/L)(8.34 lb/gal)}$$

$$\text{Number of lb, Ca(OCl)}_2 = \text{(0.0505 mil gal)(40.0 mg/L)(8.34 lb/gal)} = 16.85\text{ lb}$$

Last, determine the number of gallons needed.

$$\textbf{Number of gal, Ca(OCl)}_\mathbf{2} = \frac{\textbf{[Number of lb, Ca(OCl)}_\mathbf{2}\textbf{](100\%)}}{\textbf{(9.98 lb/gal)(Percent of Solution)}}$$

$$\text{Number of gal, Ca(OCl)}_2 = \frac{[16.85\text{ lb Ca(OCl)}_2](100\%)}{(9.98\text{ lb/gal})(60.0\%)} = \textbf{2.81 gal of Ca(OCl)}_\mathbf{2}$$

10. Solve the problem using the dilution triangle

13.0%		4.00	4.0 parts of the 13% solution are required for every 9.5 parts
	7.50%		
3.50%		5.50	5.5 parts of the 3.5% solution are required for every 9.5 parts
		9.50 total parts	

$$\frac{4.0\text{ parts(2,500 gal)}}{9.5\text{ parts}} = 1{,}053\text{ gal, round to }\textbf{1,050 gal of the 13\% solution}$$

$$\frac{5.5\text{ parts (2,500 gal)}}{9.5\text{ parts}} = \frac{1{,}447}{2{,}500\text{ gal}}\text{ gal, round to }\textbf{1,450 gal of the 3.5\% solution}$$

To make the 2,500 gal of the 7.5% solution, mix 1,050 gal of the 13% solution with 1,450 gal of the 3.5% solution.

CHAPTER 4

DISTRIBUTION

Grade 4

Students preparing for the grade 4 distribution certification test should understand all of the problems presented for grades 1, 2, and 3.

PERCENT CALCULATIONS

1. **If 12.0 lb of soda ash are mixed into 45.0 gal of water, what is the percent of soda ash in the slurry?**

   First, find the number of pounds in 45 gal of water.

   $$\text{Number of lb} = (8.34\ \text{lb/gal})(45.0\ \text{gal}) = 375.3\ \text{lb}$$

   Next, determine the percent soda ash.

   $$\text{Percent soda ash} = \frac{(12.0\ \text{lb})(100\%)}{12.0\ \text{lb} + 375.3\ \text{lb}} = \frac{(12.0\ \text{lb})(100\%)}{387.3\ \text{lb}} = \mathbf{3.10\%\ soda\ ash}$$

2. **If 3 lb, 11.5 oz of soda ash is added to 25.0 gal of water, what is the percent soda ash in the slurry?**

   First, convert ounces to pounds.

   $$\text{Number of lb} = (11.5\ \text{oz})(1\ \text{lb/16 oz}) = 0.719\ \text{lb}$$

   $$\text{Total number of lb} = 3\ \text{lb} + 0.719\ \text{lb} = 3.719\ \text{lb}$$

   $$\text{Percent soda ash} = \frac{(3.719\ \text{lb})(100\%)}{3.719\ \text{lb} + (25.0\ \text{gal})(8.34\ \text{lb/gal})} = \frac{(3.719\ \text{lb})(100\%)}{3.719\ \text{lb} + 208.5\ \text{lb}} =$$

   $$\frac{(3.719\ \text{lb})(100\%)}{212.219\ \text{lb}} = \mathbf{1.75\%\ soda\ ash}$$

3. How many pounds of lime must be added to 75 gal of water to produce a 10% lime slurry?

The equation is:

$$\textbf{Percent lime} = \frac{(x \textbf{ lb})(100\%)}{x \textbf{ lb} + (\textbf{Number of gal})(8.34 \textbf{ lb/gal})}$$

$$10\% \text{ lime} = \frac{(x \text{ lb})(100\%)}{x \text{ lb} + (75 \text{ gal})(8.34 \text{ lb/gal})}$$

Multiply both sides of the equation by [x lb + (75 gal)(8.34 lb/gal)]. This leaves the following:

$$10\% \text{ lime}[x \text{ lb} + (75 \text{ gal})(8.34 \text{ lb/gal})] = (x \text{ lb})(100\%)$$

Complete the multiplication.

$$10x \text{ \%, lb} + 6{,}255\text{\%, lb} = 100x \text{ \%, lb}$$

Subtract $10x$ %, lb from both sides of the equation.

$$6{,}255\text{\%, lb} = 90x \text{ \%, lb}$$

Last, divide both sides of the equation by 90%.

x = 69.5 lb, round to **70 lb of lime**

VOLUME DETERMINATIONS

4. Determine the volume in gallons of a trapezoidal-shaped canal that has the following dimensions:

Length = 1,240 ft | **Depth = 4.5 ft**

Bottom width = 6.0 ft (b_1) | **Top width = 10.0 ft (b_2)**

The equation is:

$$\textbf{Volume, gal} = \frac{(b_1 + b_2)(\textbf{Depth})(\textbf{Length})(7.48 \textbf{ gal/ft}^3)}{2}$$

$$\text{Volume, ft}^3 = \frac{(6.0 \text{ ft} + 10.0 \text{ ft})(4.5 \text{ ft})(1{,}240 \text{ ft})(7.48 \text{ gal/ft}^3)}{2}$$

$$\text{Volume, ft}^3 = \frac{(16.0 \text{ ft})(4.5 \text{ ft})(1{,}240 \text{ ft})(7.48 \text{ gal/ft}^3)}{2}$$

$$\text{Number of gal} = (8 \text{ ft})(4.5 \text{ ft})(1{,}240 \text{ ft})(7.48 \text{ gal/ft}^3)$$

= 333,907 gal, round to **330,000 gal**

5. A soda ash silo is conical at the bottom and cylindrical at the top. If the diameter of the cylinder is 12.5 ft with a depth of 28.0 ft, and the cone depth is 9.75 ft, what is the volume of the tank in cubic feet?

First, find the volume of the cone in cubic feet.

$$\text{Volume, ft}^3 = \tfrac{1}{3}\pi r^2(\text{Depth})$$

where the radius = Diameter/2 = 12.5 ft/2 = 6.25 ft

$$\text{Volume, ft}^3 = \tfrac{1}{3}(3.14)(6.25\text{ ft})(6.25\text{ ft})(9.75\text{ ft}) = 398.6\text{ ft}^3$$

Next, find the volume of the cylindrical part of the tank.

$$\text{Volume} = \pi r^2(\text{Depth}) = (3.14)(6.25\text{ ft})(6.25\text{ ft})(28.0\text{ ft}) = 3{,}434\text{ ft}^3$$

Last, add the two volumes for the answer.

$$\text{Total volume, ft}^3 = 398.6\text{ ft}^3 + 3{,}434\text{ ft}^3 = 3{,}833\text{ ft}^3\text{, round to } \mathbf{3{,}830\text{ ft}^3}$$

6. Determine the volume of water in gallons for the following distribution system:
Distribution pipe A is 985 ft in length and 3.0 ft in diameter.
Distribution pipe B is 645 ft in length and 2.0 ft in diameter.
The storage tank is 110 ft in diameter and has a water height of 25.36 ft.

First, find the volume in gallons for each of the distribution pipes.

The equation is:

$$\textbf{Volume, gal} = \mathbf{(0.785)(Diameter)^2(Length, ft)(7.48\ gal/ft^3)}$$

$$\text{Volume, pipe A, gal} = (0.785)(3.0\text{ ft})(3.0\text{ ft})(985\text{ ft})(7.48\text{ gal/ft}^3) = 52{,}054\text{ gal}$$

$$\text{Volume, pipe B, gal} = (0.785)(2.0\text{ ft})(2.0\text{ ft})(645\text{ ft})(7.48\text{ gal/ft}^3) = 15{,}149\text{ gal}$$

Next, determine the volume of the storage tank.
The equation is:

$$\textbf{Volume, gal} = \mathbf{(0.785)(Diameter)^2(Height, ft)(7.48\ gal/ft^3)}$$

$$\text{Volume, gal} = (0.785)(110\text{ ft})(110\text{ ft})(25.36\text{ ft})(7.48\text{ gal/ft}^3) = 1{,}801{,}797$$

Last, add the three volumes together for the total volume.

$$\text{Total volume, gal} = 52{,}054\text{ gal} + 15{,}149\text{ gal} + 1{,}801{,}797\text{ gal}$$
$$= 1{,}869{,}000\text{ gal, round to } \mathbf{1{,}900{,}000\text{ gal}}$$

DETENTION TIME CALCULATIONS

7. Calculate the theoretical detention time in hours for the following water treatment plant's clear well and associated distribution line and storage tank:

The clear well is 410.0 ft long, 99.75 ft wide, and has an average water depth of 15.5 ft. The distribution line from the clear well to the storage tank is 2.23 mi long and has a diameter of 2.5 ft. The storage tank has a diameter of 120.0 ft and averages a height of 29.8 ft of water. The flow is 28.95 mgd.

Determine the number of gallons in each of the containment structures.

The equation is:

$$\textbf{Volume, gal} = \textbf{(Length)(Width)(Depth)(7.48 gal/ft}^3\textbf{)}$$

$$\text{Volume, gal, in clear well} = (410.0 \text{ ft})(99.75 \text{ ft})(15.5 \text{ ft})(7.48 \text{ gal/ft}^3) = 4{,}741{,}656 \text{ gal}$$

Convert number of miles to feet.

$$\text{Number of ft} = (5{,}280 \text{ ft/mi})(2.23 \text{ mi}) = 11{,}774 \text{ ft}$$

The equation is:

$$\text{Volume, gal} = (0.785)(\text{Diameter})^2(\text{Length})(7.48 \text{ gal/ft}^3)$$

$$\text{Volume, gal, in pipeline} = (0.785 \text{ ft})(2.5 \text{ ft})(2.5 \text{ ft})(11{,}774 \text{ ft})(7.48 \text{ gal/ft}^3) = 432{,}091 \text{ gal}$$

$$\text{Volume, gal, in tank} = (0.785)(120.0 \text{ ft})(120.0 \text{ ft})(29.8 \text{ ft})(7.48 \text{ gal/ft}^3) = \underline{2{,}519{,}707 \text{ gal}}$$

$$7{,}693{,}454 \text{ gal}$$

Next, convert million gallons per day to gallons per hour.

$$(28.95 \text{ mgd})(1 \text{ d}/24 \text{ hr})(1{,}000{,}000/1 \text{ M}) = 1{,}206{,}250 \text{ gph}$$

Now, solve for detention time. Do not forget to write the equation with units asked for in the question.

$$\textbf{Detention time, hr} = \frac{\textbf{Volume, gal}}{\textbf{Flow rate, gph}}$$

$$\text{Detention time, hr} = \frac{7{,}693{,}454 \text{ gal}}{1{,}206{,}250 \text{ gph}} = 6.378 \text{ hr, round to } \mathbf{6.4 \text{ hr}}$$

8. Calculate the detention time to the nearest 100th hour for the following:

The clear well is 395 ft long, 103 ft wide, and has an average water depth of 11.45 ft. The distribution pipe from the clear well to the storage tank is 1.86 mi long and has a diameter of 2.0 ft. The storage tank has a diameter of 105 ft and averages a height of 28.10 ft of water. The flow is 62.07 mgd.

First, determine the number of gallons in the clear well, distribution pipe, and storage tank.

The equation is:

$$\textbf{Volume, gal} = \textbf{(Length)(Width)(Depth)(7.48 gal/ft}^3\textbf{)(Number of basins)}$$

$$\text{Volume, gal, in clear well} = (395 \text{ ft})(103 \text{ ft})(11.45 \text{ ft})(7.48 \text{ gal/ft}^3) = 3{,}484{,}508 \text{ gal}$$

Next, convert number of miles to feet.

$$\text{Number of ft} = (5{,}280 \text{ ft/mi})(1.86 \text{ mi}) = 9{,}820.8 \text{ ft}$$

$$\text{Volume, gal, in pipeline} = (0.785 \text{ ft})(2.0 \text{ ft})(2.0 \text{ ft})(9{,}820.8 \text{ ft})(7.48 \text{ gal/ft}^3) = 230{,}663 \text{ gal}$$

$$\textbf{Volume, gal} = \textbf{(0.785)(Diameter)}^2\textbf{(Height)(7.48 gal/ft}^3\textbf{)}$$

$$\text{Volume, gal in tank} = (0.785)(105 \text{ ft})(105 \text{ ft})(28.10 \text{ ft})(7.48 \text{ gal/ft}^3) = \underline{1{,}819{,}098 \text{ gal}}$$

$$5{,}534{,}269 \text{ gal}$$

Next, convert million gallons per day to gallons per hour.

$$(62.07 \text{ mgd})(1 \text{ d}/24 \text{ hr})(1{,}000{,}000/1 \text{ M}) = 2{,}586{,}250 \text{ gph}$$

Now, solve for detention time. Do not forget to write the equation with units asked for in the question.

$$\textbf{Detention time, hr} = \frac{\textbf{Volume, gal}}{\textbf{Flow rate, gph}}$$

$$\text{Detention time, hr} = \frac{5{,}534{,}269 \text{ gal}}{2{,}586{,}250 \text{ gph}} = \mathbf{2.14 \text{ hr}}$$

PRESSURE PROBLEMS

9. If the water level in a lake is 625 ft, what is the pressure in pounds per square inch at an inlet if it is 165 ft from the bottom?

First, find the depth of the inlet. It is the total depth minus the level of the inlet above the bottom.

$$\text{Depth, inlet} = 625 \text{ ft} - 165 \text{ ft} = 460 \text{ ft}$$

Next, calculate the psi.

The equation is:

$$\mathbf{psi} = \frac{\mathbf{Depth, ft}}{\mathbf{2.31\ ft/psi}}$$

$$\text{psi} = \frac{460 \text{ ft}}{2.31 \text{ ft/psi}} = \mathbf{199\ psi}$$

10. If the pressure head at a blow-off valve is 136 ft, what is the pressure in pounds per square inch?

The equation is:

$$\mathbf{Pressure\ head,\ ft} = \mathbf{(Pressure,\ psi)(2.31\ ft/psi)}$$

Rearrange the equation to solve for pressure.

$$\text{Pressure, psi} = \frac{\text{Pressure head, ft}}{2.31 \text{ ft/psi}}$$

$$\text{Pressure, psi} = \frac{136 \text{ ft}}{2.31 \text{ ft/psi}} = \mathbf{58.9\ psi}$$

11. A well has a depth of 387.4 ft. If the depth to the top of the water is 98.54 ft, what is the pressure in pounds per square inch 10.0 ft above the bottom? Disregard the additional atmospheric pressure in the well.

First, calculate the column of water in feet.

Number of ft = Total depth of well – Depth to water – Number of feet above bottom

Number of ft = 387.4 – 98.54 ft – 10.0 ft = 278.86 ft

Next, find the pressure in psi.

Pressure, psi = (Number of ft)(0.433 psi/ft) = (278.86 ft)(0.433 psi/ft)

= 120.7 psi, round to **121 psi**

12. If the pressure at the bottom of the tank is 10.8 psi, what is the height of the water in the tank?

The equation is:

Depth, ft = (Number of psi)(2.31 ft/psi)

Water level in tank = (10.8 psi)(2.31 ft/psi) = **24.9 ft**

13. If the pressure at the bottom of the lake is 26.4 psi, what is the depth of the water in the lake?

The equation is:

Depth, ft = (Number of psi)(2.31 ft/psi)

Water level in tank = (26.4 psi)(2.31 ft/psi) = **61.0 ft**

14. A tank is 70.0 ft in diameter and is 38 ft tall. If there is 890,000 gal of water in the tank, what is the psi at the bottom of the tank? What is the psi 10 ft above the bottom of the tank?

First, calculate the number of cubic feet of water present.

$$\frac{890{,}000 \text{ gal}}{7.48 \text{ gal/ft}^3} = 118{,}984 \text{ ft}^3$$

Then, calculate the number of cubic feet.

$$(0.785)(\text{Diameter})^2(\text{Depth})$$

Solve for depth.

$$118{,}984 \text{ ft}^3 = (0.785)(70 \text{ ft})(70 \text{ ft})(\text{Depth})$$

$$\text{Depth} = \frac{118{,}984 \text{ ft}^3}{(0.785)(70.0 \text{ ft})(70.0 \text{ ft})}$$

$$\text{Depth} = 30.9 \text{ ft}$$

Next, solve for the number of psi at the bottom of the tank.

$$\text{psi} = \frac{\text{Depth}}{2.31 \text{ ft/psi}} = \frac{30.9 \text{ ft}}{2.31 \text{ ft/psi}} = 13.4 \text{ psi, round to } \mathbf{13 \text{ psi}}$$

Last, solve for psi 10 ft above the bottom of the tank. The depth of the water is 10 ft less than the total depth (30.9 ft – 10 ft = 20.9 ft). Thus:

$$\text{psi} = \frac{\text{Depth}}{2.31 \text{ ft/psi}} = \frac{20.9 \text{ ft}}{2.31 \text{ ft/psi}} = \mathbf{9 \text{ psi}}$$

15. A rectangular water storage tank is partially underground. The tank is 65 ft long, 25 ft wide, and 12.5 ft high. What is the upward force of water pressure in pounds lifting up on the tank if the groundwater is 4.88 ft above the base of the tank?

First, calculate the tank's area.

The equation is:

$$\textbf{Area, ft} = \textbf{(L)(W)}$$

$$\text{Area, ft} = (65 \text{ ft})(25 \text{ ft}) = 1{,}625 \text{ ft}^2$$

Next, determine the pressure in pounds per square foot on the tank's bottom.

The equation is:

$$\textbf{Pressure, lb/ft}^2 = \textbf{(Water density, lb/ft}^3\textbf{)(Height, ft)}$$

$$\text{Pressure, lb/ft}^2 = (62.4 \text{ lb/ft}^3)(4.88 \text{ ft}) = 304.5 \text{ lb/ft}^2$$

Last, find the upward force on the bottom of the tank.

The equation is:

$$\textbf{Upward force, lb} = \textbf{(Area, ft}^2\textbf{)(Pressure, lb/ft}^2\textbf{)}$$

$$\text{Upward force, lb} = (1{,}625 \text{ ft}^2)(304.5 \text{ lb/ft}^2) = 494{,}812 \text{ lb, round to } \mathbf{490{,}000 \text{ lb}}$$

16. In problem 15, if the tank weighs 205,000 lb, what must the water level in the tank always be above in order to prevent it from floating? Assume groundwater never goes above 4.88 ft.

First, determine the pounds of water that must be in the tank by subtracting the total weight of the tank from the upward force on the tank from groundwater.

$$\text{Number of lb, water} = 494{,}812 \text{ lb} - 205{,}000 \text{ lb} = 289{,}812 \text{ lb, water}$$

Next, determine the number of gallons this would be.

$$\text{Number of gal} = 289{,}812 \text{ lb}/8.34 \text{ lb per gal} = 34{,}750 \text{ gal}$$

Then, determine the height (level) of the water in the tank.

$$\text{Level of water, ft} = \frac{34{,}750 \text{ gal}}{(1{,}625 \text{ ft}^2)(7.48 \text{ gal/ft}^3)} = 2.854 \text{ ft, round to } \mathbf{2.9 \text{ ft}}$$

17. The pressure on a 6.0-in. diameter gate valve plate is 75 psi. What is the force in pounds that is exerted on the valve plate?

First, convert the diameter in inches to feet.

$$\text{Diameter, ft} = (6\text{-in.})(1 \text{ ft}/12\text{-in.}) = 0.50 \text{ ft}$$

Then, find the cross-sectional area of the gate valve in square feet.

$$\text{Area} = (0.785)(\text{Diameter})^2$$

$$\text{Area} = (0.785)(0.50 \text{ ft})(0.50 \text{ ft})$$

$$\text{Area} = 0.196 \text{ ft}^2$$

Now, find the force in pounds.

The equation is:

$$\textbf{Force, lb} = \textbf{(Area, ft}^2\textbf{)(144 in.}^2\textbf{/ft}^2\textbf{)(Number psi)}$$

Substituting:

$$\text{Force, lb} = (0.196 \text{ ft}^2)(144 \text{ in.}^2/\text{ft}^2)(75 \text{ psi}) = 2{,}116.8 \text{ lb, round to } \mathbf{2{,}100 \text{ lb}}$$

18. Determine the head loss in pounds per square inch for a pipe that has a water velocity of 5 ft/s.

First, find the head loss in feet.

The equation is:

$$\textbf{Head loss, ft} = \frac{\textbf{(Velocity)}^2}{\textbf{2(g)}}$$

where g is the acceleration of gravity, which is 32.2 ft/s^2.

$$\text{Head loss, ft} = \frac{(5 \text{ ft/s})^2}{2(32.2 \text{ ft/s}^2)} = 0.388 \text{ ft}$$

Next, convert head loss in feet to head loss in psi.

$$\text{Head loss, psi} = (0.388 \text{ ft})(0.433 \text{ psi/ft}) = \mathbf{0.17 \text{ psi}}$$

METER FLOW DETERMINATIONS

19. What should the flow meter read in gallons per minute if a 12-in. diameter main is to be flushed at 4.9 ft/s?

First, convert the diameter of the pipe from inches to feet.

$$\text{Number of ft} = (12 \text{ in.})(1 \text{ ft}/12 \text{ in.}) = 1.0 \text{ ft}$$

Next, calculate the pipe's cross-sectional area in square feet.

$$\textbf{Area, ft}^2 = \textbf{(0.785)(Diameter)}^2$$

$$\text{Area, ft}^2 = (0.785)(1.0 \text{ ft})(1.0 \text{ ft}) = 0.785 \text{ ft}^2$$

Next, find the flow in the pipe in cubic feet per second.

$$\textbf{Flow, ft}^3\textbf{/s} = \textbf{(Area, ft}^2\textbf{)(Velocity, ft/s)}$$

$$\text{Flow, ft}^3\text{/s} = (0.785 \text{ ft}^2)(4.9 \text{ ft/s}) = 3.846 \text{ ft}^3\text{/s}$$

Last, determine the reading on the flow meter in gallons per minute.

Flow, gpm = (3.846 ft^3/s)(7.48 gal/ft^3)(60 s/min) = 1,726 gpm, round to **1,700 gpm**

20. What should the flow meter read in gallons per minute if an 18.0-in. diameter main is to be flushed at 4.75 ft/s?

First, convert the diameter of the pipe from inches to feet.

$$\text{Number of ft} = (18.0 \text{ in.})(1 \text{ ft}/12 \text{ in.}) = 1.50 \text{ ft}$$

Then, calculate the pipe's cross-sectional area in square feet.

$$\textbf{Area, ft}^2 = \textbf{(0.785)(Diameter)}^2$$

$$\text{Area, ft}^2 = (0.785)(1.50 \text{ ft})(1.50 \text{ ft}) = 1.766 \text{ ft}^2$$

Next, find the flow in the pipe in cubic feet per second.

$$\textbf{Flow, ft}^3\textbf{/s} = \textbf{(Area, ft}^2\textbf{)(Velocity, ft/s)}$$

$$\text{Flow, ft}^3\text{/s} = (1.766 \text{ ft}^2)(4.75 \text{ ft/s}) = 8.388 \text{ ft}^3\text{/s}$$

Last, determine the reading on the flow meter in gallons per minute.

Flow, gpm = (8.388 ft^3/s)(7.48 gal/ft^3)(60 s/min) = 3,765 gpm, round to **3,760 gpm**

21. If a pump discharges 8,240 gal in 1 hr, what will it discharge in 15 hr and 15 min?

First, convert the number of minutes to hours.

$$\text{Number of hr} = 15 \text{ hr} + (15 \text{ min})(1 \text{ hr}/60 \text{ min}) = 15 \text{ hr} + 0.25 \text{ hr} = 15.25 \text{ hr}$$

$$\text{Total gal discharged} = (8{,}240 \text{ gal/hr})(15.25 \text{ hr}) = 125{,}660 \text{ gal, round to } \mathbf{126{,}000 \text{ gal}}$$

FLOW PROBLEMS

22. The velocity through a channel is 1.88 ft/s. If the channel is 9.45 ft wide and 3.1 ft deep, what is the flow in cubic feet per second? Assume the channel is basically square.

The equation is:

$$\mathbf{Q = (A)(V)} \text{ or } \mathbf{Flow, ft^3/s = (Area, ft^2)(Velocity, ft/s)}$$

$$\text{Q, ft}^3\text{/s} = (1.88 \text{ ft/s})(9.45 \text{ ft})(3.1 \text{ ft}) = 55.07 \text{ ft}^3\text{/s, round to } \mathbf{55 \text{ ft}^3/s}$$

23. Determine the flow in feet per second if a water flow of 677 gpm is going through an 8.0-in. pipe.

First, convert the number of gallons per minute to cubic feet per second.

$$\text{Number of ft}^3\text{/s} = \frac{677 \text{ gpm}}{(7.48 \text{ gal/ft}^3)(60 \text{ s/min})} = 1.5 \text{ ft}^3\text{/s}$$

Next, convert the diameter from inches to feet.

$$\text{Number of ft} = (8.0 \text{ in.})(1 \text{ ft}/12 \text{ in.}) = 0.667 \text{ ft}$$

The equation for flow is:

$$\mathbf{Flow, ft^3/s = (Area, ft^2)(Velocity, ft/s)}$$

where the area $= (0.785)(\text{Diameter})^2$

$$1.5 \text{ ft}^3\text{/s} = (0.785)(0.667 \text{ ft})(0.667 \text{ ft})(\text{Flow, ft/s})$$

Rearrange the problem to solve for flow.

$$\text{Flow, ft/s} = \frac{1.5 \text{ ft}^3\text{/s}}{(0.785)(0.667 \text{ ft})(0.667 \text{ ft})} = 4.295 \text{ ft/s, round to } \mathbf{4.3 \text{ ft/s}}$$

24. Find the instantaneous flow in cubic feet per second for a trapezoidal channel that is 4.6 ft wide at the bottom and 7.6 ft wide at the water surface. The depth of the water is 49 in. and the flow velocity is 1.95 ft/s.

First, convert the depth of the water from inches to feet.

49 in./12 in. per foot = 4.08 ft in depth

The equation is:

$$\textbf{Flow, ft}^3\textbf{/s} = \frac{\textbf{(Bottom width) + (Water surface width)(Depth, ft)(Velocity, ft/s)}}{\textbf{2}}$$

$$\text{Flow, ft}^3\text{/s} = \frac{(4.6 \text{ ft} + 7.6 \text{ ft})(4.08 \text{ ft})(1.95 \text{ ft/s})}{2} = 48.53 \text{ ft}^3\text{/s, round to } \mathbf{49\ ft^3/s}$$

25. Water is flowing at a velocity of 2.63 ft/s in an 8.0-in. diameter pipe. If the pipe changes from the 8-in. to a 14-in. pipe, what will the velocity be in the 14-in. pipe?

The flow in the 8.0-in. pipe equals the flow in the 14-in. pipe, as the flow must remain constant.

$$\mathbf{Q_1 = Q_2}$$

Because Q, flow = (Area)(Velocity), it follows that:

(Area 1)(Velocity 1) = (Area 2)(Velocity 2)

First, find the diameters in feet for the 8.0-in. and 14-in. pipes.

Diameter for 8.0-in. = 8.0-in.(1 ft/12-in.) = 0.667 ft

Diameter for 14-in. = 14-in.(1 ft/12-in.) = 1.167 ft

Then, determine the areas of each pipe.

$$\textbf{Area} = \mathbf{(0.785)(Diameter)^2}$$

$$\text{Area 1 (8.0-in.)} = (0.785)(0.667 \text{ ft})(0.667 \text{ ft}) = 0.349 \text{ ft}^2$$

$$\text{Area 2 (14-in.)} = (0.785)(1.167 \text{ ft})(1.167 \text{ ft}) = 1.069 \text{ ft}^2$$

Last, substitute areas calculated and known velocity in the 8.0-in. pipe.

$$(0.349 \text{ ft}^2)(2.63 \text{ ft/s}) = (1.069 \text{ ft}^2)(x, \text{ ft/s})$$

Solve for x.

$$x, \text{ ft/s} = \frac{(0.349 \text{ ft}^2)(2.63 \text{ ft/s})}{(1.069 \text{ ft}^2)} = 0.859 \text{ ft/s, round to } \textbf{0.86 ft/s in 14-in. pipe}$$

26. What is the flow (velocity) through a 4.0-in. diameter pipe if it is delivering 175 gpm?

First, convert the diameter of the pipe to feet.

$$(4.0 \text{ in.})(1 \text{ ft}/12 \text{ in.}) = 0.333 \text{ ft}$$

Next, determine the cross-sectional area of the pipe.

The equation is:

$$\textbf{Area, ft}^2 = \textbf{(0.785)(Diameter)}^2$$

$$\text{Area, ft}^2 = (0.785)(0.333 \text{ ft})(0.333 \text{ ft}) = 0.087 \text{ ft}^2$$

Then, convert gallons per minute to cubic feet per second.

$$\text{Number of ft}^3/\text{s} = \frac{175 \text{ gpm}}{(7.48 \text{ gal/ft}^3)(60 \text{ s/min})} = 0.390 \text{ ft}^3/\text{s}$$

Last, determine the flow in feet per second.

The equation is:

$$\text{Flow, ft}^3/\text{s} = (\text{Area, ft}^2)(\text{Velocity, ft/s})$$

$$\text{Velocity, ft/s} = \frac{(0.390 \text{ ft}^3/\text{s})}{(0.087 \text{ ft}^2)} = 4.48 \text{ ft/s, round to } \textbf{4.5 ft/s}$$

27. Water is flowing at a velocity of 3.95 ft/s in an 6.0-in. diameter pipe. If the pipe changes from the 6.0-in. to a 10.0-in. pipe, what will the velocity be in the 10.0-in. pipe?

The flow in the 6.0-in. pipe equals the flow in the 10.0-in. pipe, as the flow must remain constant.

$$\mathbf{Q_1 = Q_2}$$

Because Q, flow = (Area)(Velocity), it follows that:

$$\textbf{(Area 1)(Velocity 1) = (Area 2)(Velocity 2)}$$

First, find the diameters in feet for the 6.0-in. and 10-in. pipes.

$$\text{Diameter for 6.0-in.} = 6.0\text{-in.}(1 \text{ ft}/12\text{-in.}) = 0.5 \text{ ft}$$

$$\text{Diameter for 10.0-in.} = 10.0\text{-in.}(1 \text{ ft}/12\text{-in.}) = 0.833 \text{ ft}$$

Then, determine the areas of each pipe.

$$\text{Area} = (0.785)(\text{Diameter})^2$$

$$\text{Area 1 (6.0-in.)} = (0.785)(0.5\ \text{ft})(0.5\ \text{ft}) = 0.196\ \text{ft}^2$$

$$\text{Area 2 (10.0-in.)} = (0.785)(0.833\ \text{ft})(0.833\ \text{ft}) = 0.545\ \text{ft}^2$$

Last, substitute areas calculated and known velocity in the 6.0-in. pipe.

$$(0.196\ \text{ft}^2)(3.95\ \text{ft/s}) = (0.545\ \text{ft}^2)(x,\ \text{ft/s})$$

Solve for x.

$$x,\ \text{ft/s} = \frac{(0.196\ \text{ft}^2)(3.95\ \text{ft/s})}{(0.545\ \text{ft}^2)} = 1.42\ \text{ft/s, round to } \mathbf{1.4\ ft/s\ in\ 10.0\text{-}in.\ pipe}$$

28. Water is flowing at a velocity of 1.28 ft/s in an 14-in. diameter pipe. If the pipe changes from the 14-in. to an 8.0-in. pipe, what will the velocity be in the 8.0-in. pipe?

The flow in the 14-in. pipe equals the flow in the 8.0-in. pipe, as the flow must remain constant.

$$\mathbf{Q_1 = Q_2}$$

Because Q, flow = (Area)(Velocity), it follows that:

$$\mathbf{(Area\ 1)(Velocity\ 1) = (Area\ 2)(Velocity\ 2)}$$

First, find the diameters in feet for the 14-in. and 8.0-in. pipes.

$$\text{Diameter for 14-in.} = \text{14-in.}(1\ \text{ft/12-in.}) = 1.167\ \text{ft}$$

$$\text{Diameter for 8.0-in.} = \text{8.0-in.}(1\ \text{ft/12-in.}) = 0.667\ \text{ft}$$

Then, determine the areas of each pipe.

$$\text{Area} = (0.785)(\text{Diameter})^2$$

$$\text{Area 1 (14-in.)} = (0.785)(1.167\ \text{ft})(1.167\ \text{ft}) = 1.069\ \text{ft}^2$$

$$\text{Area 2 (8.0-in.)} = (0.785)(0.667\ \text{ft})(0.667\ \text{ft}) = 0.349\ \text{ft}^2$$

Last, substitute areas calculated and known velocity in the 14-in. pipe.

$$(1.069\ \text{ft}^2)(1.28\ \text{ft/s}) = (0.349\ \text{ft}^2)(x,\ \text{ft/s})$$

Solve for x.

$$x,\ \text{ft/s} = \frac{(1.069\ \text{ft}^2)(1.28\ \text{ft/s})}{(0.349\ \text{ft}^2)} = 3.92\ \text{ft/s, round to } \mathbf{3.9\ ft/s\ in\ 8.0\text{-}in.\ pipe}$$

29. Determine the amount of gallons that were used from a storage tank for a particular day in question, given the following data:

Diameter of tank = 100.0 ft

Initial water level at beginning of day = 32.56 ft

Final water level at end of day = 28.33 ft

Water pumped to tank = 802 gpm (average)

First, determine the amount of water in the storage tank at the beginning (initial) of the day.

The equation is:

$$\textbf{Volume, gal} = \mathbf{(0.785)(Diameter)^2(Height, ft)(7.48\ gal/ft^3)}$$

$$\text{Volume, gal} = (0.785)(100.0\ \text{ft})(100.0\ \text{ft})(32.56\ \text{ft})(7.48\ \text{gal/ft}^3) = 1{,}911{,}858\ \text{gal}$$

Next, use the same equation to find the number of gallons left (final) in the tank at the end of the day.

$$\text{Volume, gal} = (0.785)(100.0\ \text{ft})(100.0\ \text{ft})(28.33\ \text{ft})(7.48\ \text{gal/ft}^3) = 1{,}663{,}481\ \text{gal}$$

Then, find the number of gallons that were pumped to the tank that day.

$$\text{Number of gal} = (\text{Pumping rate, gpm})(1{,}440\ \text{min/d})$$

$$\text{Number of gal} = (802\ \text{gpm})(1{,}440\ \text{min/d}) = 1{,}154{,}880\ \text{gal}$$

Now, determine the gallons actually used that day.

$$\text{Number of gal used} = \text{Initial gallon} - \text{Final gallon} + \text{Pumped water}$$

$$\text{Number of gal used} = 1{,}911{,}858 - 1{,}663{,}481 + 1{,}154{,}880$$

$$= 1{,}403{,}257\ \text{gal, round to } \mathbf{1{,}400{,}000\ gal}$$

FLUSHING TIME PROBLEMS

30. A 12-in. main line needs to be flushed. How many minutes will it take to flush the line at 30 gpm if the desired length of pipeline to be flushed is 200 ft?

First, convert the pipe diameter from inches to feet.

$$\text{Number of ft} = (12\ \text{in.})(1\ \text{ft}/12\ \text{in.}) = 1.0\ \text{ft}$$

Next, determine the volume in gallons for 200 ft of the 12-in. pipeline.

The equation is:

$$\textbf{Volume, gal} = \textbf{(0.785)(Diameter)}^2\textbf{(Length)(7.48 gal/ft}^3\textbf{)}$$

$$\text{Volume, gal} = (0.785)(1.0)(1.0)(200 \text{ ft})(7.48 \text{ gal/ft}^3) = 1{,}174 \text{ gal}$$

Then, find the flushing time.

$$\text{Flushing time, min} = \frac{1{,}174 \text{ gal}}{30 \text{ gpm}} = 39.13\text{, round to } \textbf{40 min}$$

31. A 31-ft, 1.0-in. service line requires flushing. How many minutes are required to flush the line if the line is flushed at a rate of 12 gpm and 25 volumes are removed?

First, convert the pipe diameter from inches to feet.

$$\text{Number of ft} = (1.0 \text{ in.})(1 \text{ ft/12 in.}) = 0.0833 \text{ ft}$$

Next, determine the volume in gallons for 31 ft of the 1.0-in. pipeline.

The equation is:

$$\textbf{Volume, gal} = \textbf{(0.785)(Diameter)}^2\textbf{(Length)(7.48 gal/ft}^3\textbf{)(Number of volumes removed)}$$

$$\text{Volume, gal} = (0.785)(0.0833 \text{ ft})(0.0833 \text{ ft})(31 \text{ ft})(7.48 \text{ gal/ft}^3)(25 \text{ volumes}) = 31.58 \text{ gal}$$

Then, find the flushing time.

$$\text{Flushing time, min} = \frac{31.58 \text{ gal}}{12 \text{ gpm}} = \textbf{2.6 min}$$

32. A 14-in. main line needs to be flushed. If a 100.0-ft section of the pipeline was flushed for 28 min, what was the flushing rate in gallons per minute?

First, convert the pipe diameter from inches to feet.

$$\text{Number of ft} = (14 \text{ in.})(1 \text{ ft/12 in.}) = 1.167 \text{ ft}$$

Next, determine the volume in gallons for 100.0 ft of the 14-in. pipeline.

(continued)

The equation is:

$$\textbf{Volume, gal} = \mathbf{(0.785)(Diameter)^2(Length)(7.48\ gal/ft^3)}$$

$$\text{Volume, gal} = (0.785)(1.167\ \text{ft})(1.167\ \text{ft})(100.0\ \text{ft})(7.48\ \text{gal/ft}^3)$$

$$\text{Volume, gal} = 799.67\ \text{gal}$$

Then, find the flushing rate in gallons per minute.

$$\text{Gallons removed} = \frac{799.67\ \text{gal}}{28\ \text{min}} = 28.56\ \text{gpm, round to } \mathbf{29\ gpm}$$

33. A distribution pipe that is 36-in. in diameter delivers 17,600,000 gal in 24 hr. What is the average flow during the 24 hr time period in feet per second?

First, convert the diameter in inches to feet.

$$\text{Number of ft} = (36\ \text{in.})(1\ \text{ft}/12\ \text{in.}) = 3.0$$

Next, determine the cross-sectional area of the pipe.

The equation is:

$$\textbf{Area, ft}^2 = \mathbf{(0.785)(Diameter)^2}$$

$$\text{Area, ft}^2 = (0.785)(3.0\ \text{ft})(3.0\ \text{ft}) = 7.065\ \text{ft}$$

Next, determine the cubic feet per second.

$$\text{Number of ft}^3\text{/s} = (17{,}600{,}000\ \text{gal}/24\ \text{hr})(1\ \text{ft}^3/7.48\ \text{gal})(1\ \text{hr}/3{,}600\ \text{s})$$

$$= 27.23\ \text{ft}^3\text{/s}$$

Last, determine the flow in feet per second.

The equation is:

$$\textbf{Flow, ft}^3\textbf{/s} = \mathbf{(Area,\ ft^2)(Velocity,\ ft/s)}$$

Rearrange the equation to solve for velocity.

$$\text{Average velocity, ft/s} = \frac{\text{Flow, ft}^3\text{/s}}{\text{Area, ft}^2}$$

Substituting:

$$\text{Average velocity, ft/s} = \frac{27.23\ \text{ft}^3\text{/s}}{7.065\ \text{ft}^2} = 3.85\ \text{ft/s, round to } \mathbf{3.9\ ft/s}$$

DENSITY CALCULATIONS

34. What is the density of a liquid alum solution in grams per cubic centimeter if the alum solution weighs 21.34 lb and occupies 2.00 gal?

First, convert the number of pounds to grams.

The following applies:

$$1 \text{ lb} = 454 \text{ g and } 1 \text{ L} = 1{,}000.027 \text{ cm}^3$$

$$\textbf{Number of g} = \textbf{(Number of lb)(454 g/lb)}$$

$$\text{Number of g} = (21.34 \text{ lb})(454 \text{ g/lb}) = 9{,}688 \text{ g}$$

Next, convert the number of gallons to liters.

$$\text{Number of L} = (2.00 \text{ gal})(3.785 \text{ L/gal}) = 7.57 \text{ L}$$

$$\text{Number of cm}^3 = (1{,}000.027 \text{ cm}^3/1 \text{ L})(7.57 \text{ L}) = 7{,}570 \text{ cm}^3$$

$$\text{Density} = \text{Mass/Volume}$$

$$\text{Density} = 9{,}688 \text{ g}/7570 \text{ cm}^3 = \mathbf{1.28 \text{ g/cm}^3}$$

35. A solution weighs 22.5 lb and occupies a space of 8.00 L. What is its density in pounds per gallon?

First, convert the number of liters to gallons.

$$\text{Number of gal} = (\text{Number of liters})(1 \text{ gal}/3.785 \text{ L})$$

$$\text{Number of gal} = (8.00 \text{ L})(1 \text{ gal}/3.785 \text{ L}) = 2.11 \text{ gal}$$

$$\text{Density} = \text{Mass/Volume}$$

In this case, weight is substituted for mass.

$$\text{Density} = 22.5 \text{ lb}/2.11 \text{ gal} = \mathbf{10.7 \text{ lb/gal}}$$

SPECIFIC GRAVITY OF LIQUIDS

Specific gravity (sp gr) is a dimensionless number that compares the density of one substance to that of another. Water is the standard for liquids with a specific gravity (sp gr) of 1.0. The density of water can be expressed in several ways: 62.4 lb/ft^3, 8.34 lb/gal, or 1 g/cm^3.

36. How many pounds per gallon will a solution weigh if it has a specific gravity of 1.24?

The equation is:

Specific gravity = Density of substance/Density of water

Rearrange the equation to solve for the density of the liquid in pounds per cubic foot.

$$\text{Density of liquid, lb/ft}^3 = (\text{Specific gravity})(\text{Density of water, lb gal})$$

$$\text{Density of liquid, lb/ft}^3 = (1.24)(8.34\text{ lb/gal}) = 10.34\text{ lb/ft}^3\text{, round to } \mathbf{10.3\ lb/ft^3}$$

37. What is the specific gravity of a polymer solution that weighs 10.85 lb/gal?

The equation is:

Specific gravity = Density of unknown substance/Density of water

$$\text{Specific gravity of polymer solution} = \frac{10.85\text{ lb/gal}}{8.34\text{ lb/gal}} = \mathbf{1.30\ sp\ gr}$$

RATIO PROBLEMS

38. If 8.0 gal of a calcium hypochlorite solution are used to treat 100,000 gal of water, how many gallons of calcium hypochlorite will be needed to treat 525,000 gal assuming the same concentration is desired?

Set up a ratio.

Let x = unknown (number of gallons of hypochlorite needed).

The equation is:

$$\frac{\boldsymbol{x}\ \textbf{lb, Ca(OCl)}_2}{\textbf{Number of gal}_2} = \frac{\textbf{Number of lb, Ca(OCl)}_2}{\textbf{Number of gal}_1}$$

$$\frac{x \text{ gal, Ca(OCl)}_2}{525{,}000 \text{ gal}} = \frac{8.0 \text{ gal, Ca(OCl)}_2}{100{,}000 \text{ gal}}$$

$$x \text{ lb, Ca(OCl)}_2 = \frac{(8.0 \text{ gal, Ca(OCl)}_2(525{,}000 \text{ gal})}{100{,}000 \text{ gal}} = \textbf{42 gal of calcium hypochlorite}$$

39. If a pump moves 1,245,500 gal in exactly 32 hr, how much will it move in exactly 24 hr if all other conditions remain the same?

Let x = the number of gallons pumped in 24 hr.

$$\frac{x\text{, gal pumped}}{24 \text{ hr}} = \frac{1{,}245{,}500 \text{ gal}}{32 \text{ hr}}$$

$$x\text{, gal pumped} = \frac{(1{,}245{,}500 \text{ gal})(24 \text{ hr})}{32 \text{ hr}} = 934{,}125 \text{ gal, round to } \textbf{934,120 gal pumped}$$

40. A treatment plant is chlorinating a flow of 34.9 ft³/s at a rate 215 lb/d. If the flow is changed to 41.4 ft³/s, what should be the new chlorinator setting in pounds per day?

Let x = unknown (pounds per day, Cl_2).

The equation is:

$$\frac{\mathbf{x\ lb/d,\ Cl_2}}{\mathbf{Flow\ Rate_2,\ ft^3/s}} = \frac{\mathbf{Number\ of\ lb/d\ Cl_2}}{\mathbf{Flow\ Rate_1,\ ft^3/s}}$$

$$\frac{x \text{ lb/d, } Cl_2}{41.4 \text{ ft}^3\text{/s}} = \frac{215 \text{ lb/d } Cl_2}{34.9 \text{ ft}^3\text{/s}}$$

$$x \text{ lb/d, } Cl_2 = \frac{(41.4 \text{ ft}^3\text{/s})(215 \text{ lb/d } Cl_2)}{34.9 \text{ ft}^3\text{/s}} = \mathbf{255\ lb/d\ of\ Cl_2}$$

DOSAGE PROBLEMS

41. A water treatment plant is treating 22.3 mgd with 315 lb/d of chlorine. What is the chlorine dosage in milligrams per liter?

The equation is:

$$\textbf{Pounds per day, chlorine} = \textbf{(mgd)(Dosage, mg/L)(8.34 lb/gal)}$$

$$315 \text{ lb/d} = (22.3 \text{ mgd})(\text{Dosage, mg/L})(8.34 \text{ lb/gal})$$

Solve for dosage.

$$\text{Dosage, mg/L} = \frac{315 \text{ lb/d}}{(22.3 \text{ mgd})(8.34 \text{ lb/gal})} = \textbf{1.69 mg/L of chlorine}$$

42. A storage tank that is going to be put back into service requires disinfection at a dosage of 30 mg/L. If the tank has a diameter of 102 ft and is 28.1 ft in height at the overflow, how many gallons of a 10.25% sodium hypochlorite solution will be needed if the tank is filled to 10.0% capacity?

First, determine the capacity in gallons of the storage tank.

The equation is:

$$\textbf{Volume, gal} = \textbf{(0.785)(Diameter)}^2\textbf{(Height)(7.48 gal/ft}^3\textbf{)}$$

$$\text{Volume, gal} = (0.785)(102\text{ ft})(102\text{ ft})(28.1\text{ ft})(7.48\text{ gal/ft}^3)(10.0\%/100\%) = 171{,}663\text{ gal}$$

Next, convert gallons to million gallons.

$$\text{Number of mil gal} = (171{,}663\text{ gal})(1\text{ M}/1{,}000{,}000) = 0.17166\text{ mil gal}$$

Next, find the number of pounds of sodium hypochlorite needed using the "pounds" equation.

Note: Drop the "day" on both sides of the equation as it is not needed.

The equation is:

$$\textbf{Pounds of chlorine} = \textbf{(mil gal)(Dosage, mg/L)(8.34 lb/gal)}$$

$$\text{Number of lb, chlorine} = (0.17166\text{ mil gal})(30\text{ mg/L})(8.34\text{ lb/gal}) = 42.95\text{ lb}$$

$$\text{Number of gal, NaOCl} = \frac{(\text{Chlorine, lb})(100\%)}{(8.34\text{ lb/gal})(\text{Hypochlorite, }\%)}$$

$$\text{Number of gal, NaOCl} = \frac{(42.95\text{ lb})(100\%)}{(8.34\text{ lb/gal})(10.25)} = 50.2\text{ gal, round to } \mathbf{50\text{ gal of NaOCl}}$$

Note: They should use calcium hypochlorite as it has higher available chlorine. Thus, substantially fewer gallons will be needed.

43. A 62.5% calcium hypochlorite solution was used to treat 15.8 mil gal. The tank containing the hypochlorite solution is 10.0 ft in diameter. If the tank dropped 1.19 in. during the time the 15.8 mil gal were treated, what must have been the chlorine dosage in milligrams per liter?

First, convert the level drop in the tank from inches to feet.

$$\text{Number of ft} = (1.19 \text{ in.})(1 \text{ ft}/12 \text{ in.}) = 0.0992 \text{ ft}$$

Next, determine the amount of hypochlorite used.

The equation is:

$$\textbf{Volume, gal} = \textbf{(0.785)(Diameter of tank)}^2\textbf{(Level drop, ft)(7.48 gal/ft}^3\textbf{)}$$

$$\text{Volume, gal} = (0.785)(10.0 \text{ ft})(10.0 \text{ ft})(0.0992 \text{ ft})(7.48 \text{ gal/ft}^3) = 58.13 \text{ gal}$$

Next, find how many pounds of available chlorine this is.

$$\text{Number of lb, chlorine} = \frac{(58.13 \text{ gal})(8.34 \text{ lb/gal})(62.5\%)}{100\%} = 303 \text{ lb, chlorine}$$

Last, calculate the dosage.

The equation is:

$$\textbf{Number of pounds} = \textbf{(mil gal)(Dosage, mg/L)(8.34 lb/gal)}$$

Solve for dosage.

$$\text{Dosage, mg/L} = \frac{\text{Number of lb, chlorine}}{(\text{mil gal})(8.34 \text{ lb/gal})}$$

$$\text{Dosage, mg/L} = \frac{303 \text{ lb, chlorine}}{(15.8 \text{ mil gal})(8.34 \text{ lb/gal})} = \textbf{2.30 mg/L of calcium hypochlorite}$$

44. A well that is 345 ft in depth and 12 in. in diameter requires disinfection. Depth to the water from the top of the casing is 129 ft. If the desired dose is 50.0 mg/L, what is the number of gallons of calcium hypochlorite (60.5% available chlorine) required?

First, find the length (in feet) of water in the casing.

Length of water-filled casing = Depth of well – Depth of water to top of casing

Length of water-filled casing = 345 ft – 129 ft = 216 ft

Then, convert the diameter from inches to feet.

$$\text{Diameter, ft} = \frac{12 \text{ in.}}{12 \text{ in./ft}} = 1.0 \text{ ft}$$

Next, determine the volume in gallons of water in the well casing.

The equation is:

$$\textbf{Volume, in gal} = \mathbf{(0.785)(Diameter)^2(Length)(7.48\ gal/ft^3)}$$

$$\text{Volume, in gal} = (0.785)(1.0 \text{ ft})(1.0 \text{ ft})(216 \text{ ft})(7.48 \text{ gals/ft}^3) = 1{,}268 \text{ gal}$$

Next, determine the number of million gallons.

Million gallons = (1,268 gal)(1 M/1,000,000) = 0.001268 mil gal

Using the “pounds” equation, calculate the number of pounds of chlorine needed.

Pounds of chlorine = (0.001268 mil gal)(50.0 mg/L)(8.34 lb/gal) = 0.5288 lb

Last, calculate the gallons of calcium hypochlorite solution required.

$$\textbf{Number of gal, } \mathbf{Ca(OCl)_2} = \frac{\textbf{(Chlorine, lb)(100\%)}}{\textbf{(8.34 lb/gal)(Hypochlorite, \%)}}$$

$$\text{Number of gal, } Ca(OCl)_2 = \frac{(0.5288 \text{ lb})(100\%)}{(8.34 \text{ lb/gal})(60.5\%)}$$

= 0.1048 gal, round to **0.10 gal of $Ca(OCl)_2$**

45. What must have been the chlorine dose if 125 gal of a 10.0% sodium hypochlorite solution was used to treat 250,000 gal of water?

First, determine the number of pounds of chlorine used.

The equation is:

$$\textbf{Chlorine, lb} = \frac{\textbf{(Hypochlorite, gal)(Hypochlorite, \%)(8.34 lb/gal)}}{\textbf{100\%}}$$

$$\text{Chlorine, lb} = \frac{(125 \text{ gal})(10.0\%)(8.34 \text{ lb/gal})}{100\%} = 104.25 \text{ lb}$$

Next, convert gallons of water to million gallons.

$$\text{Number mil gal} = (250{,}000 \text{ gal})(1 \text{ M}/1{,}000{,}000) = 0.25 \text{ mil gal}$$

Last, determine the actual chlorine dosage.

The equation is:

$$\textbf{Dosage, mg/L} = \frac{\textbf{Pounds of chlorine}}{\textbf{(Volume, mil gal)(8.34 lb/gal)}}$$

$$\text{Dosage, mg/L} = \frac{104.25 \text{ lb}}{(0.25 \text{ mil gal})(8.34 \text{ lb/gal})} = \mathbf{50 \text{ mg/L}}$$

46. A dosage of 0.12 mg/L of copper sulfate pentahydrate is desired in a reservoir containing trout. (Trout are sensitive to copper sulfate, and it has been shown that dosages over 0.14 mg/L are toxic to this fish.) The capacity of the reservoir is 4,500 acre-ft. If the available copper is 25%, how many pounds of copper sulfate pentahydrate are needed?

First, convert the volume of the reservoir to million gallons.

$$(43{,}560 \text{ ft}^3\text{/acre-ft})(4{,}500 \text{ acre-ft})(7.48 \text{ gal/ft}^3)(1 \text{ M}/1{,}000{,}000) = 1{,}466 \text{ mil gal}$$

The equation is:

$$\textbf{Copper sulfate, lb} = \frac{\textbf{(mil gal)(Dosage, mg/L)(8.34 lb/gal)}}{\textbf{Percent available copper/100\%}}$$

$$\text{Copper sulfate, lb} = \frac{(1{,}466 \text{ mil gal})(0.12 \text{ mg/L})(8.34 \text{ lb/gal})}{25\%/100\%}$$

Copper sulfate, lb = 5,869 lb, round to **5,900 lb of copper sulfate**

47. A pipe that is 3.00 ft in diameter and 675 ft long is to be disinfected with a 63.5% calcium hypochlorite solution. If the desired dose is 35.0 mg/L, what is the number of calcium hypochlorite gallons needed?

First, calculate the number of gallons in the pipeline.

The equation is:

$$\textbf{Number of gal} = \textbf{(0.785)(Diameter, ft)}^2\textbf{(Length, ft)(7.48 gal/ft}^3\textbf{)}$$

$$\text{Number of gal} = (0.785)(3.00\text{ ft})(3.00\text{ ft})(675\text{ ft})(7.48\text{ gal/ft}^3) = 35{,}671\text{ gal}$$

Convert gallons to million gallons.

$$\text{Number of mil gal} = (35{,}671\text{ gal})(1\text{ M}/1{,}000{,}000) = 0.03567\text{ mil gal}$$

Next, determine the number of pounds of calcium hypochlorite needed using the "pounds" equation.

$$\text{Number of lb, } Ca(OCl)_2 = (\text{mil gal})(\text{Dosage, mg/L})(8.34\text{ lb/gal})$$

$$\text{Number of lb, } Ca(OCl)_2 = (0.03567\text{ mil gal})(35.0\text{ mg/L})(8.34\text{ lb/gal}) = 10.41\text{ lb}$$

Last, determine the number of gallons needed.

$$\textbf{Number of gal, } \mathbf{Ca(OCl)_2} = \frac{\textbf{[Number of lb, } \mathbf{Ca(OCl)_2}\textbf{](100\%)}}{\textbf{(8.34 lb/gal)(Percent of solution)}}$$

$$\text{Number of gal, } Ca(OCl)_2 = \frac{[10.41\ Ca(OCl)_2](100\%)}{(8.34\text{ lb/gal})(63.5\%)} = \mathbf{1.97\ gal\ of\ Ca(OCl)_2}$$

48. A treatment plant is processing water at a rate of 12.8 ft³/s. If lime is being added at a rate of 122.85 g/min, what are the lime usage in pounds per day and the dosage in milligrams per liter?

First, find the pounds per day of lime usage.

The equation is:

$$\text{Pounds per day, lime} = (\text{g/min})(1{,}440\text{ min/d})(1\text{ lb}/454\text{ g}) = \text{lb/d}$$

$$\text{Pounds per day, lime} = (122.85\text{ g/min})(1{,}440\text{ min/d})(1\text{lb}/454\text{ g})$$

Pounds per day, lime = 389.66 lb/d, round to **390 lb/d of lime**

(continued)

Next, convert 12.8 ft^3/s to the number of million gallons per day.

Number of mgd = $(12.8\ ft^3/s)(86{,}400\ s/d)(7.48\ gal/ft^3)(1\ M/1{,}000{,}000) = 8.27$ mgd

Rearrange and solve for dosage in milligrams per liter.

$$\text{Dosage, mg/L} = \frac{389.66\ \text{lb/d}}{(8.27\ \text{mgd})(8.34\ \text{lb/gal})} = \mathbf{5.65\ mg/L\ of\ lime}$$

49. A plant uses a coagulant aid that weighs 10.76 lb/gal to help treat 7.22 mgd. The result of a drawdown test is 182 mL for the coagulant aid in 5 min. What is the coagulant aid dosage in milligrams per liter?

First, determine the number of ml/min for the coagulant aid.

$$\frac{182\ \text{mL}}{5\ \text{min}} = 36.4\ \text{mL/min}$$

Use the dosage equation.

$$\text{Milligrams per liter, coagulant aid} = \frac{(\text{mL/min})(1{,}440\ \text{min/d})(\text{lb/gal, coagulant aid})}{(3{,}785\ \text{mL/gal})(\text{mgd})(8.34\ \text{lb/gal})}$$

$$\text{Milligrams per liter, coagulant aid} = \frac{(36.4\ \text{mL/min})(1{,}440\ \text{min/d})(10.76\ \text{lb/gal})}{(3{,}785\ \text{mL/gal})(7.22\ \text{mgd})(8.34\ \text{lb/gal})}$$

Milligrams per liter, coagulant aid = **2.47 mg/L of coagulant aid**

50. A tank that is 6.5 ft in diameter and 10.25 ft high at the overflow requires disinfection. How much 12.5% sodium hypochlorite will be required for a dosage of 50.0 mg/L?

First, determine the capacity of the largest tank in gallons.

The equation is:

$$\mathbf{Volume,\ gal = (0.785)(Diameter)^2(Height,\ ft)(7.48\ gal/ft^3)}$$

$$\text{Volume, gal} = (0.785)(6.5\ \text{ft})(6.5\ \text{ft})(10.25\ \text{ft})(7.48\ \text{gal/ft}^3) = 2{,}543\ \text{gal}$$

Next, convert number of gallons to million gallons.

$$(2{,}543 \text{ gal})(1 \text{ M}/1{,}000{,}000) = 0.002543 \text{ mil gal}$$

Using the "pounds" equation, determine the chlorine pounds needed.

$$\text{Number of pounds} = (\text{Volume, mil gal})(\text{Dosage, mg/L})(8.34 \text{ lb/gal})$$

$$\text{Number of pounds} = (0.002543 \text{ mil gal})(50.0 \text{ mg/L})(8.34 \text{ lb/gal}) = 1.06 \text{ lb}$$

$$\textbf{NaOCl solution, gal} = \frac{\textbf{(Chlorine, lb)(100\%)}}{\textbf{(8.34 lb/gal)(Hypochlorite, \%)}}$$

$$\text{NaOCl solution, gal} = \frac{(1.06 \text{ lb})(100\%)}{(8.34 \text{ lb/gal})(12.5\%)} = 1.017 \text{ gal, round to } \textbf{1.0 gal of NaOCl}$$

51. A storage facility has two tanks. The largest is 100.0 ft in diameter and is 34.1 ft high at the overflow. The smaller tank is 58.0 ft in diameter and is 25.3 ft high at the overflow. The tanks are to be disinfected with an initial dose of 100.0 mg/L. How many gallons of 62.5% calcium hypochlorite solution will be required for each tank if they are filled to 10.0% capacity?

First, determine the capacity of the largest tank in gallons.

The equation is:

$$\textbf{Volume, gal} = \textbf{(0.785)(Diameter)}^2\textbf{(Height, ft)(7.48 gal/ft}^3\textbf{)}$$

$$\text{Volume, gal} = (0.785)(100.0 \text{ ft})(100.0 \text{ ft})(34.1 \text{ ft})(7.48 \text{ gal/ft}^3)(10.0\%/100\%) = 200{,}228 \text{ gal}$$

Next, convert number of gallons to million gallons.

$$(200{,}228 \text{ gal})(1 \text{ M}/1{,}000{,}000) = 0.2002 \text{ mil gal}$$

Using the "pounds" equation, determine the chlorine pounds needed.

$$\text{Number of pounds} = (\text{Volume, mil gal})(\text{Dosage, mg/L})(8.34 \text{ lb/gal})$$

$$\text{Number of pounds} = (0.2002 \text{ mil gal})(100.0 \text{ mg/L})(8.34 \text{ lb/gal}) = 167 \text{ lb}$$

$$\textbf{Ca(OCl)}_2 \textbf{ solution, gal} = \frac{\textbf{(Chlorine, lb)(100\%)}}{\textbf{(8.34 lb/gal)(Hypochlorite, \%)}}$$

(continued)

$$\text{Ca(OCl)}_2\text{ solution, gal} = \frac{(167\text{ lb})(100\%)}{(8.34\text{ lb/gal})(62.5\%)} = \mathbf{32.0\text{ gal of Ca(OCl)}_2\text{ needed}}$$

Now, determine the number of gallons required for the smaller tank.

$$\text{Volume, gal} = (0.785)(58.0\text{ ft})(58.0\text{ ft})(25.3\text{ ft})(7.48\text{ gal/ft}^3)(10.0\%/100\%) = 49{,}974\text{ gal}$$

Next, convert number of gallons to million gallons.

$$(49{,}974\text{ gal})(1\text{ M}/1{,}000{,}000) = 0.04997\text{ mil gal}$$

Using the "pounds" equation, determine the chlorine pounds needed.

$$\text{Number of lb} = (\text{Volume, mil gal})(\text{Dosage, mg/L})(8.34\text{ lb/gal})$$

$$\text{Number of lb} = (0.04997\text{ mil gal})(100\text{ mg/L})(8.34\text{ lb/gal}) = 41.7\text{ lb}$$

$$\mathbf{Ca(OCl)_2\text{ solution, gal} = \frac{(\text{Chlorine, lb})(100\%)}{(8.34\text{ lb/gal})(\text{Hypochlorite, }\%)}}$$

$$\text{Ca(OCl)}_2\text{ solution, gal} = \frac{(41.7\text{ lb})(100\%)}{(8.34\text{ lb/gal})(62.5\%)} = \mathbf{8.00\text{ gal of Ca(OCl)}_2}$$

52. How many calcium hypochlorite tablets each weighing 0.35 lb will be needed to disinfect a water main given the following information:

Length of pipe = 1,670 ft	**Pipe diameter = 3.0 ft**
Calcium hypochlorite = 63.2% available chlorine	**Dosage required = 30.0 mg/L**

First, determine the volume of the pipeline in million gallons.

$$\mathbf{\text{Number of mil gal} = \frac{(0.785)(\text{Diameter})^2(\text{Length, ft})(7.48\text{ gal/ft}^3)}{1{,}000{,}000/\text{M}}}$$

$$\text{Number of mil gal} = \frac{(0.785)(3.0\text{ ft})(3.0\text{ ft})(1{,}670\text{ ft})(7.48\text{ gal/ft}^3)}{1{,}000{,}000/\text{M}} = 0.088\text{ mil gal}$$

Next, determine the number of pounds of calcium hypochlorite needed.

$$\text{Number of lb, Ca(OCl)}_2 = \frac{(\text{mil gal})(\text{Dosage, mg/L})(8.34\text{ lb/gal})}{\text{Percent purity}}$$

$$\text{Number of lb, } Ca(OCl)_2 = \frac{(0.088\text{ mil gal})(30.0\text{ mg/L})(8.34\text{ lb/gal})}{63.2\%/100\%} = 34.84\text{ lb}$$

Last, find the number of tablets needed.

$$\text{Number of tablets} = \frac{34.84\text{ lb}}{0.35\text{ lb per tablet}} = 99.5\text{ tablets, round to } \mathbf{100\text{ tablets of calcium hypochlorite}}$$

53. A 25.0% calcium hypochlorite solution needs to be made up in a mobile tank in order to disinfect a remote storage tank. The supervisor does not want any 25.0% calcium hypochlorite solution left in the mobile tank so we must only make what is needed. Given the following data, determine the amount of 65.0% calcium hypochlorite solution required for the 25.0% solution and the number of gallons of the 25.0% solution needed to disinfect the tank:

Small tank diameter = 4.95 ft	**Height of small tank = 4.20 ft**
Storage tank diameter = 44.9 ft	**Height of overflow = 19.5 ft**
Dosage required = 25.0 mg/L	

First, we must determine the number of gallons of the 25.0% solution needed to disinfect the tank. Then, we can determine how much of the 65.0% solution will be needed to make up the 25.0% solution.

Determine the number of gallons in the storage tank when full.

The equation is:

$$\textbf{Number of gal} = \mathbf{(0.785)(Diameter)^2(Height, ft)(7.48\ gal/ft^3)}$$

$$\text{Number of gal} = (0.785)(44.9\text{ ft})(44.9\text{ ft})(19.5\text{ ft})(7.48\text{ gal/ft}^3) = 230{,}833\text{ gal}$$

Next, convert gallons to million gallons.

$$\text{Number of mil gal} = \frac{230{,}833\text{ gal}}{1{,}000{,}000/1\text{ M}} = 0.2308\text{ mil gal}$$

Next, find the number of pounds of calcium hypochlorite needed.

The equation is:

$$\textbf{Number of lb} = \mathbf{(mil\ gal)(Dosage, mg/L)(8.34\ lb/gal)}$$

$$\text{Number of lb} = (0.2308\text{ mil gal})(25.0\text{ mg/L})(8.34\text{ lb/gal}) = 48.12\text{ lb}$$

(continued)

Next, calculate the number of gallons of a 25.0% solution of calcium hypochlorite.

Number of gal, $Ca(OCl)_2$ =

$$\frac{48.12 \text{ lb}}{(8.34 \text{ lb/gal})(25.0\%/100\%)} = \textbf{23.1 gal of the 25.0\% Ca(OCl)}_2 \textbf{ solution}$$

Solve for the amount of 65.0% calcium hypochlorite solution needed by using the dilution triangle.

Note: the 65.0% calcium hypochlorite solution will be mixed with water to make the 25.0% solution, hence the zero in the dilution triangle below.

65.0%		25.0*	25.0 parts of the 65.0% solution is required for every 65 parts
	25.0%		
0%		40.0†	40.0 parts of water is required for every 65 parts
		65 total parts	

$$\frac{25.0 \text{ parts } (23.1 \text{ gal})}{65 \text{ parts}} = \textbf{8.9 gal of the 65.0\% solution}$$

$$\frac{40.0 \text{ parts } (23.1 \text{ gal})}{65 \text{ parts}} = \textbf{14.2 gal of the water}$$

23.1 gal (added as a crosscheck of the math)

To make the 23.1 gal of the 25.0% solution, mix 8.9 gal of the 65.0% solution with 14.2 gal of water.

* This number is calculated by subtracting diagonally 25% from 0%, absolute (negative sign is dropped).
† This number is calculated by subtracting diagonally 25% from 65%.

54. How many calcium hypochlorite tablets will be needed to disinfect a water main given the following information:

Length of pipe A = 765 ft	**Pipe A diameter = 2.50 ft**
Length of pipe B = 1,220 ft	**Pipe B diameter = 1.50 ft**
Calcium hypochlorite = 61.5% available chlorine	**Dosage required = 50.0 mg/L**
Each tablet is 0.38 lb.	

First, determine the volume of the pipeline in million gallons.

$$\textbf{Number of mil gal} = \frac{\textbf{(0.785)(Diameter)}^2\textbf{(Length, ft)(7.48 gal/ft}^3\textbf{)}}{\textbf{1,000,000/M}}$$

$$\text{Number of mil gal, pipe A} = \frac{(0.785)(2.50 \text{ ft})(2.50 \text{ ft})(765 \text{ ft})(7.48 \text{ gal/ft}^3)}{1{,}000{,}000/\text{M}} = 0.02807 \text{ mil gal}$$

$$\text{Number of mil gal, pipe B} = \frac{(0.785)(1.5 \text{ ft})(1.5 \text{ ft})(1{,}220 \text{ ft})(7.48 \text{ gal/ft}^3)}{1{,}000{,}000/\text{M}} = 0.01612 \text{ mil gal}$$

Next, add the number of million gallons in the two pipes.

$$\text{Total number of mil gal} = 0.02807 \text{ mil gal} + 0.01612 \text{ mil gal} = 0.04419 \text{ mil gal}$$

Next, determine the number of pounds of calcium hypochlorite needed.

$$\text{Number of lb, Ca(OCl)}_2 = \frac{(\text{mil gal})(\text{Dosage, mg/L})(8.34 \text{ lb/gal})}{\text{Percent purity}}$$

$$\text{Number of lb, Ca(OCl)}_2 = \frac{(0.04419 \text{ mil gal})(50.0 \text{ mg/L})(8.34 \text{ lb/gal})}{61.5\% \text{ available chlorine}/100\%} = 29.96 \text{ lb}$$

Last, find the number of tablets needed.

$$\text{Number of tablets} = \frac{29.96 \text{ lb}}{0.38 \text{ lb per tablet}} = 78.84, \text{ round to } \mathbf{79 \text{ tablets}}$$

55. A water treatment plant has a filter flow of 2,950 gpm and is being treated with 435 gpd of a hypochlorite solution. If the desired dose is 2.5 mg/L, determine the concentration of the hypochlorite solution in percent.

First, convert 2,950 gpm filter flow to million gallons per day.

$$\frac{(2{,}950 \text{ gpm})(1{,}440 \text{ min})}{\text{day}} \frac{(1 \text{ mil gal})}{1{,}000{,}000 \text{ gal}} = 4.248 \text{ mgd}$$

Then, convert 435 gpd of the hypochlorite solution rate of flow to million gallons per day.

$$\frac{(435 \text{ gpd})\ (1 \text{ mil gal})}{1{,}000{,}000 \text{ gal}} = 0.000435 \text{ mgd}$$

Then, use the equal dosage equations.

$$(0.000435 \text{ mgd})(x \text{ mg/L})(8.34 \text{ lb/gal}) = (4.248 \text{ mgd})(2.5 \text{ mg/L})(8.34 \text{ lb/gal})$$

$$\text{mg/L, hypochlorite} = \frac{(4.248 \text{ mgd})(2.5 \text{ mg/L})(8.34 \text{ lb/gal})}{(0.000435 \text{ mgd})(8.34 \text{ lb/gal})} = 24{,}414 \text{ mg/L}$$

(continued)

Last, convert milligrams per liter into percent.

$$\frac{(24{,}414\ \text{mg/L})(1\%)}{10{,}000\ \text{mg/L}} = \mathbf{2.4\%}$$

56. A plant treats 23.5 mgd with a cationic polymer that weighs 10.36 lb/gal. If a drawdown test gives a result of 375 mL in 5 min, what is the polymer dosage in milligrams per liter?

First, determine the number of milliliters per minute for the polymer.

$$\frac{(375\ \text{mL})}{5\ \text{min}} = 75.0\ \text{mL/min}$$

The dosage/conversion equation is:

$$\text{mg/L, polymer} = \frac{(\text{mL/min})(1{,}440\ \text{min/d})(\text{lb/gal, polymer})}{(3{,}785\ \text{mL/gal})(\text{mgd})(8.34\ \text{lb/gal})}$$

$$\text{mg/L, polymer} = \frac{(75.0\ \text{mL/min})(1{,}440\ \text{min/d})(10.36\ \text{lb/gal})}{(3{,}785\ \text{mL/gal})(23.5\ \text{mgd})(8.34\ \text{lb/gal})} = \mathbf{1.51\ mg/L\ of\ polymer}$$

57. A pipe that is 2.50 ft in diameter and 1,560 ft long is to be disinfected with 59.5% calcium hypochlorite solution. If the desired dose is 50.0 mg/L, what is the number of calcium hypochlorite gallons needed?

First, calculate the number of gallons in the pipeline.

The equation is:

$$\textbf{Number of gal} = \mathbf{(0.785)(Diameter, ft)^2(Length, ft)(7.48\ gal/ft^3)}$$

$$\text{Number of gal} = (0.785)(2.50\ \text{ft})(2.50\ \text{ft})(1{,}560\ \text{ft})(7.48\ \text{gal/ft}^3) = 57{,}250\ \text{gal}$$

Then, convert gallons to million gallons.

$$\text{Number of mil gal} = (57{,}250\ \text{gal})(1\ \text{M}/1{,}000{,}000) = 0.05725\ \text{mil gal}$$

Next, determine the number of pounds of calcium hypochlorite needed using the "pounds" equation.

$$\text{Number of lb, Ca(OCl)}_2 = \text{(mil gal)(Dosage, mg/L)(8.34 lb/gal)}$$

$$\text{Number of lb, Ca(OCl)}_2 = \text{(0.05725 mil gal)(50.0 mg/L)(8.34 lb/gal)}$$

$$= \text{23.87 lb of Ca(OCl)}_2$$

Last, determine the number of gallons needed.

$$\textbf{Number of gal, Ca(OCl)}_2 = \frac{\textbf{[Number of lb, Ca(OCl)}_2\textbf{](100\%)}}{\textbf{(8.34 lb/gal)(Percent of solution)}}$$

$$\text{Number of gal, Ca(OCl)}_2 = \frac{\text{[23.87 lb Ca(OCl)}_2\text{](100\%)}}{\text{(8.34 lb/gal)(59.5\%)}} = \textbf{4.81 gal of Ca(OCl)}_2$$

58. A well that is 348 ft in depth and 16.0 in. in diameter requires disinfection. Depth to the water from the top of the casing is 119 ft. If the desired dose is 50.0 mg/L, what is the number of pounds and ounces of calcium hypochlorite (62.5% available chlorine) required? Assume the calcium hypochlorite is 9.50 lb/gal.

First, find the length (in feet) of water in the casing.

$$\text{Length of water-filled casing} = \text{Depth of well} - \text{Depth of water to top of casing}$$

$$\text{Length of water-filled casing} = 348 \text{ ft} - 119 \text{ ft} = 229 \text{ ft}$$

Then, convert the diameter from inches to feet.

$$\text{Diameter, ft} = \frac{16.0 \text{ in.}}{12 \text{ in./ft}} = 1.33 \text{ ft}$$

Next, determine the volume in gallons of water in the well casing.

$$\textbf{Volume, in gal} = \textbf{(0.785)(Diameter)}^2\textbf{(Length)(7.48 gal/ft}^3\textbf{)}$$

$$\text{Volume, in gal} = (0.785)(1.33 \text{ ft})(1.33 \text{ ft})(229 \text{ ft})(7.48 \text{ gal/ft}^3) = 2{,}379 \text{ gal}$$

Next, determine the number of million gallons.

$$\text{Million gallons} = (2{,}379 \text{ gal})(1 \text{ M}/1{,}000{,}000) = 0.002379 \text{ mil gal}$$

(continued)

Last, using the "pounds" equation, calculate the number of pounds of calcium hypochlorite.

$$\text{Pounds, calcium hypochlorite} = \frac{(0.002379 \text{ mil gal})(50.0 \text{ mg/L})(8.34 \text{ lb/gal})}{(62.5\% \text{ available chlorine}/100\%)} = 1.587 \text{ lb}$$

Next, find the number of ounces.

$$\text{Number of oz} = \frac{(1.587 \text{ lb})(128 \text{ oz/gal})}{9.50 \text{ lb/gal}}$$

= 21.38 oz, round to **21.4 oz of calcium hypochlorite**

59. Soda ash was added to a mixing tank at an average rate of 37 g/min. The soda ash solution was then injected into the effluent pipe of the distribution storage tank (clear well). This tank was 120 ft in diameter. If the average pumping rate into the tank over the past 24 hr was 3,110 gpm and the level in the tank increased by 2.05 ft, what was the average dosage in milligrams per liter?

In order to solve this problem we need to determine how many gallons of water was treated with the soda ash and the number of pounds of soda ash used for this treatment.

Begin by finding the number of gallons pumped to the storage tank.

$$\text{Number of gallons pumped} = (3{,}110 \text{ gpm})(1{,}440 \text{ min}) = 4{,}478{,}400 \text{ gal}$$

Next, determine the gallons from the 2.05 ft increase in the tank's water level.

The equation is:

$$\text{Number of gal} = (0.785)(\text{Diameter})^2(\text{Level difference, ft})(7.48 \text{ gal/ft}^3)$$

$$\text{Number of gal} = (0.785)(120 \text{ ft})(120 \text{ ft})(2.05 \text{ ft})(7.48 \text{ gal/ft}^3) = 173{,}336 \text{ gal}$$

Because the level in the tank increased, more water was pumped into the tank than left the tank. Therefore, to find the number of gallons treated by the soda ash in 24 hr or 1 day, subtract the number of gallons pumped by the increase in gallons in the tank.

$$\text{Total water treated, gal} = 4{,}478{,}400 \text{ gal} - 173{,}336 \text{ gal} = 4{,}305{,}064 \text{ gal}$$

Next, convert gallons to million gallons.

$$\text{Number of mil gal} = \frac{4{,}305{,}064 \text{ gal}}{1{,}000{,}000/1 \text{ M}} = 4.305 \text{ mil gal}$$

Now, determine the soda ash usage in pounds.

Number of lb = (37 g/min)(1,440 min/d)(1 lb/454 g) = 117.36 lb of soda ash

Last, determine the dosage using the "pounds" equation.

Number of lb, soda ash = (mgd)(Dosage, mg/L)(8.34 lb/gal)

Rearrange and solve for dosage in milligrams per liter.

$$\text{Dosage, mg/L} = \frac{117.36 \text{ lb/d}}{(4.305 \text{ mgd})(8.34 \text{ lb/gal})} = \textbf{3.3 mg/L of soda ash}$$

60. Calculate the chlorine dosage in milligrams per liter given the following data:
The sodium hypochlorite solution has 7.50% available chlorine.
The diameter of the hypochlorite tank is 8.0 ft.
The level in the hypochlorite tank dropped 6.96 in. in 24 hr.
The water treated in that 24 hr was 6.52 mil gal.

First, convert the level drop in the tank from inches to feet.

Number of ft = (6.96 in.)(1 ft/12 in.) = 0.58 ft

Next, determine the amount of hypochlorite used.

The equation is:

Volume, gal = (0.785)(Diameter of Tank)^2(Level Drop, ft)(7.48 gal/ft^3)

Volume, gal = (0.785)(8.0 ft)(8.0 ft)(0.58 ft)(7.48 gal/ft^3) = 217.96 gal

Next, find how many pounds of available chlorine this is.

$$\text{Number of lb, chlorine} = \frac{(217.96 \text{ gal})(8.34 \text{ lb/gal})(7.50\%)}{100\%} = 136.33 \text{ lb, chlorine}$$

Last, calculate the dosage.

Number of lb = (mil gal)(Dosage, mg/L)(8.34 lb/gal)

(continued)

Solve for dosage.

$$\text{Dosage, mg/L} = \frac{\text{Number of lb, chlorine}}{\text{(mil gal)(8.34 lb/gal)}}$$

$$\text{Dosage, mg/L} = \frac{\text{136.33 lb, chlorine}}{\text{(6.52 mil gal)(8.34 lb/gal)}} = \textbf{2.51 mg/L of chlorine}$$

61. Soda ash slurry is being added to water that is being released from a storage tank to the distribution system to raise the pH. If the amount of soda ash being added averages 12.72 g/min for that day, and the water leaving the distribution system averages 675 gpm for that day, what must have been the soda ash dosage in milligrams per liter?

First, determine the amount of water released from the storage tank in million gallons per day.

Number of mgd = (675 gpm)(1,440 min/d)(1 M/1,000,000) = 0.972 mgd

Next, find the number of pounds of soda ash used for the day.

Number of lb, soda ash = (12.72 g/min)(1,440 min/d)(1 lb/454 g) = 40.35 lb, soda ash

Last, calculate the dosage.

The equation is:

Number of lb = (mil gal)(Dosage, mg/L)(8.34 lb/gal)

Solve for dosage.

$$\text{Dosage, mg/L} = \frac{\text{Number of lb, soda ash}}{\text{(mil gal)(8.34 lb/gal)}}$$

$$\text{Dosage, mg/L} = \frac{\text{40.35 lb, soda ash}}{\text{(0.972 mil gal)(8.34 lb/gal)}} = \textbf{4.98 mg/L}$$

CORROSION PROBLEMS

62. Test results of distribution water give a pH of 7.65 and a pH_s of 7.73. What is the Langelier Index and what does this tell you about the distribution system in the area where it was collected?

The equation is:

$$\text{Langelier Index} = \text{pH} - \text{pH}_s$$

$$\text{LI} = 7.65 - 7.73 = \mathbf{0.08}$$

Note: A negative LI indicates the water is corrosive.

63. Test results of distribution water give a pH of 7.78 and a pH_s of 7.64. What is the Langelier Index and what does this tell you about the distribution system in the area where it was collected?

The equation is:

$$\text{Langelier Index} = \text{pH} - \text{pH}_s$$

$$\text{LI} = 7.78 - 7.64 = \mathbf{0.14}$$

Note: A positive LI indicates the water is scale forming.

SPECIFIC YIELD AND DRAWDOWN CALCULATIONS

64. If a well produces 237 gpm with a drawdown of 18.4 ft, what is the specific yield in gallons per minute per foot?

The equation is:

$$\textbf{Specific yield, gpm/ft} = \frac{\textbf{Well yield, gpm}}{\textbf{Drawdown, ft}}$$

$$\text{Specific yield, gpm/ft} = \frac{237 \text{ gpm}}{18.4 \text{ ft}} = 12.88 \text{ gpm/ft, round to } \mathbf{12.9 \text{ gpm/ft}}$$

65. Find the drawdown of a well if the well yields 312 gpm and the specific yield is 18.6.

Write the equation, arranging it to solve for the unknown, drawdown.

The equation is:

$$\textbf{Drawdown, ft} = \frac{\textbf{Well yield, gpm}}{\textbf{Specific yield, gpm/ft}}$$

$$\text{Drawdown, ft} = \frac{312 \text{ gpm}}{18.6 \text{ gpm/ft}} = 16.774 \text{ ft, round to } \mathbf{16.8 \text{ ft}}$$

66. Calculate the specific yield in gallons per minute per foot for a well that produces an average of 189 gpm and has a drawdown of 15.7 ft.

The equation is:

$$\textbf{Specific yield, gpm/ft} = \frac{\textbf{Well yield, gpm}}{\textbf{Drawdown, ft}}$$

$$\text{Specific yield, gpm/ft} = \frac{189 \text{ gpm}}{15.7 \text{ ft}} = 12.04 \text{ gpm/ft, round to } \mathbf{12.0 \text{ gpm/ft}}$$

PERCENT MIXTURE PROBLEMS

67. How many gallons of a 3.4% solution must be mixed with a 10.5% solution to make exactly 500 gal of a 6.2% solution?

Solve the problem using the dilution triangle.

10.5%		2.8*	2.8 parts of the 10.5% solution are required for every 7.1 parts
	6.2%		
3.4%		4.3†	4.3 parts of the 3.4% solution are required for every 7.1 parts
		7.1 total parts	

$\frac{2.8 \text{ parts } (500 \text{ gal})}{7.1 \text{ parts}} =$ **197 gal of the 10.5% solution**

$\frac{4.3 \text{ parts } (500 \text{ gal})}{7.1 \text{ parts}} =$ **303 gal of the 3.4% solution**

500 gal (added as a cross check of math)

To make the 500 gal of the 6.2% solution, mix 197 gal of the 10.5% solution with 303 gal of the 3.4% solution.

* This number is calculated by subtracting diagonally 6.2% from 3.4%, absolute (negative sign is dropped).
† This number is calculated by subtracting diagonally 6.2% from 10.5%.

68. A solution containing exactly 400 gal of a 5.0% sodium hypochlorite solution is required. How many gallons of a 12.5% solution must be mixed with a 2.20% solution to make the required solution?

Solve the problem using the dilution triangle.

12.5%		2.8	2.8 parts of the 12.5% solution are required for every 10.3 parts
	5.0%		
2.20%		7.5	7.5 parts of the 2.20% solution are required for every 10.3 parts
		10.3 total parts	

$\frac{2.8 \text{ parts } (400 \text{ gal})}{10.3 \text{ parts}} =$ **109 gal of the 12.5% solution**

$\frac{7.5 \text{ parts } (400 \text{ gal})}{10.3 \text{ parts}} =$ **291 gal of the 2.20% solution**

400 gal

To make the 400 gal of the 5.0% sodium hypochlorite solution, mix 109 gal of the 12.5% solution with 291 gal of the 2.20% solution.

69. A solution containing exactly 245 gal of 7.5% hypochlorite is required. How many gallons of a 12.4% solution must be mixed with a 3.0% solution to make the required solution? Give the answer to the nearest gallon.

Solve the problem using the dilution triangle.

12.4%		4.5	4.5 parts of the 12.4% solution are required for every 9.4 parts
	7.5%		
3.0%		4.9	4.9 parts of the 3.0% solution are required for every 9.4 parts
		9.4 total parts	

$$\frac{4.5 \text{ parts}(245 \text{ gal})}{9.4 \text{ parts}} = \mathbf{117 \text{ gal of the } 12.4\% \text{ solution}}$$

$$\frac{4.9 \text{ parts}(245 \text{ gal})}{9.4 \text{ parts}} = \mathbf{128 \text{ gal of the } 3.0\% \text{ solution}}$$

245 gal

To make the 245 gal of the 7.5% solution, mix 117 gal of the 12.4% solution with 128 gal of the 3.0% solution.

70. If 91 lb of high test hypochlorite (HTH) with 60.25% available chlorine are mixed with 100 gal of water, what percent is the calcium hypochlorite solution? What percent solution of 100% available chlorine is this?

The equation is:

$$\textbf{Percent HTH solution} = \frac{\textbf{(lb HTH)(100\%)}}{\textbf{(Number of gal)(8.34 lb/gal)}}$$

$$\text{Percent HTH solution} = \frac{(91 \text{ lb HTH})(100\%)}{(100 \text{ gal})(8.34 \text{ lb/gal})} = 10.91\%, \text{ round to } \mathbf{11\% \text{ HTH solution}}$$

$$\text{Percent available chlorine solution} = (10.91\%)(60.25\%/100\%)$$

$$= \mathbf{6.6\% \text{ available chlorine}}$$

71. How many pounds of high test hypochlorite are needed to make exactly 400 gal of a 12.0% high test hypochlorite solution?

The equation is:

$$\textbf{Percent HTH solution} = \frac{\textbf{(lb HTH)(100\%)}}{\textbf{(Number of gal)(8.34 lb/gal)}}$$

Rearrange the equation.

$$\text{Number of lb, HTH} = \text{(Percent solution)(Number of gal)(8.34 lb/gal)/100\%}$$

$$\text{Number of lb, HTH} = \text{(12.0\% solution)(400 gal)(8.34 lb/gal)/100\%} = \textbf{400 lb of HTH}$$

72. How many pounds of 100% lime must be added to 75 gal of water to produce a 5% lime slurry?

Let x = the number of pounds of lime needed.

$$\text{Percent lime} = \frac{(x\text{ lb})(100\%)}{(x\text{ lb}) + (8.34\text{ lb/gal})(\text{Number of gal})}$$

In order to solve this problem, we will have to use a little more complex algebra then what we have been using so far. It is not that difficult, just follow the steps and always remember that whatever we do to one side of the equation, we have to do to the other side.

First, multiply both sides by (x lb) + (8.34 lb/gal)(75 gal).

$$(x\text{ lb})(5.0\%\text{ lime}) + (8.34\text{ lb/gal})(75\text{ gal})(5.0\%\text{ lime}) = \frac{(x\text{ lb})(100\%)[(x\text{ lb}) + (8.34\text{ lb/gal})(75\text{ gal})]}{[(x\text{ lb}) + (8.34\text{ lb/gal})(75\text{ gal})]}$$

The terms in the [ ] will cancel each other in the numerator (top) and the denominator (bottom) to get the following:

$$(x\text{ lb})(5.0\%\text{ lime}) + (8.34\text{ lb/gal})(75\text{ gal})(5.0\%\text{ lime}) = (x\text{ lb})(100\%)$$

Next, divide both sides by 100%.

$$\frac{(x\text{ lb})(5.0\%\text{ lime})}{100\%} + \frac{(8.34\text{ lb/gal})(75\text{ gal})(5.0\%\text{ lime})}{100\%} = \frac{(x\text{ lb})(100\%)}{100\%}$$

(continued)

The 100% will cancel on the right side of the equation to get the following:

$$\frac{(x \text{ lb})(5.0\% \text{ lime})}{100\%} + \frac{(8.34 \text{ lb/gal})(75 \text{ gal})(5.0\% \text{ lime})}{100\%} = x \text{ lbs}$$

Next, reduce the problem by solving the multiplication and division of the following:

$$(8.34 \text{ lb/gal})(75 \text{ gal})(5.0\% \text{ lime})/100\% = 31.275 \text{ lb}$$

and the division of:

$$(x \text{ lb})(5.0\% \text{ lime})/100\%$$

which gives:

$$0.05x \text{ lb of lime} + 31.275 \text{ lb of lime} = x \text{ lb}$$

Next, subtract 0.05 lb of lime from both sides to get the following:

$$0.05x \text{ lb of lime} - 0.05x \text{ lb of lime} + 31.275 \text{ lb of lime} = x \text{ lb} - 0.05x \text{ lb of lime}$$

Add the like terms (the x terms) to get the following:

$$31.275 \text{ lb} = 0.95x \text{ lb of lime}$$

Last, divide both sides by 0.95.

$$\frac{31.275 \text{ lb}}{0.95} = \frac{0.95x \text{ lb of lime}}{0.95}$$

x lb of lime = 32.92 lb of lime, round to **33 lb of lime**

DETERMINING PERCENT STRENGTH OR CONCENTRATION OF A SOLUTION

73. What is the percent strength of a solution if 380 gal of a 15.8% solution is mixed with 119 gal of a 2.50% solution? The 15.8% solution has a specific gravity of 1.23 and the 2.5% solution weighs 9.28 lb/gal.

First, find the weight in pounds per gallon for the 15.8% solution.

$$\text{Weight of 15.8\% solution} = (\text{Specific gravity})(8.34 \text{ lb/gal})$$

$$\text{Weight of 15.8\% solution} = (1.23 \text{ sp gr})(8.34 \text{ lb/gal}) = 10.26 \text{ lb/gal}$$

The equation is:

$$\textbf{Percent mixture} = \frac{\textbf{Soln. 1, gal(lb/gal)(Avail. \%/100\%) + Soln. 2, gal(lb/gal)(Avail. \%/100\%)}}{\textbf{Soln. 1, gal(lb/gal) + Soln. 2, gal(lb/gal)}} \times \textbf{100 strength}$$

$$\text{Percent mixture strength} = \frac{380 \text{ gal}(10.26 \text{ lb/gal})(15.8\%/100\%) + 119 \text{ gal}(9.28 \text{ lb/gal})(2.5\%/100\%)}{380 \text{ gal}(10.26 \text{ lb/gal}) + 119 \text{ gal}(9.28 \text{ lb/gal})} \times 100\%$$

$$\text{Percent mixture strength} = \frac{(616 \text{ lb} + 27.61 \text{ lb})(100\%)}{3{,}898.8 \text{ lb} + 1{,}104.32 \text{ lb}} = \frac{(643.61 \text{ lb})(100\%)}{5{,}003.12 \text{ lb}}$$

= 12.86%, round to **13% strength of new solution**

74. A water treatment plant has a filter flow of 2,875 gpm and is being treated with 280 gpd of a hypochlorite solution. If the desired dose is 2.0 mg/L, determine the concentration of the hypochlorite solution in percent.

First, convert 2,875 gpm filter flow to million gallons per day.

$$\frac{(2{,}875 \text{ gpm})(1{,}440 \text{ min})\ (1 \text{ mil gal})}{\text{day} \quad 1{,}000{,}000 \text{ gal}} = 4.14 \text{ mgd}$$

Then, convert 280 gpd of the hypochlorite solution rate of flow to million gallons per day.

$$\frac{(280 \text{ gpd})\ (1 \text{ mil gal})}{1{,}000{,}000 \text{ gal}} = 0.00028 \text{ mgd}$$

(continued)

Then, use the equal dosage equations.

$$\mathbf{(0.00028\ mgd)(\mathit{x}\ mg/L)(8.34\ lb/gal) = (4.14\ mgd)(2.0\ mg/L)(8.34\ lb/gal)}$$

$$\text{mg/L, hypochlorite} = \frac{(4.14\text{ mgd})(2.0\text{ mg/L})(8.34\text{ lb/gal})}{(0.00028\text{ mgd})(8.34\text{ lb/gal})} = 29{,}571\text{ mg/L}$$

Last, convert milligrams per liter into percent.

$$\frac{(29{,}571\text{ mg/L})\quad(1\%)}{10{,}000\text{ mg/L}} = \mathbf{3.0\%\ of\ hypochlorite}$$

75. What is the percent strength of a mixture of two zinc orthophosphate (ZOP) solutions given the following parameters?

ZOP solution 1: 685 gal 62.5% solution and a specific gravity of 1.36

ZOP solution 2: 1,488 gal of a 78% solution and a specific gravity of 1.29

First, determine the weight in pounds per gallon for both of the solutions.

The equation is:

$$\mathbf{Weight\ of\ solution = (Specific\ gravity)(8.34\ lb/gal)}$$

$$\text{Weight of solution 1} = (1.36\text{ sp gr})(8.34\text{ lb/gal}) = 11.34\text{ lb/gal}$$

$$\text{Weight of solution 2} = (1.29\text{ sp gr})(8.34\text{ lb/gal}) = 10.76\text{ lb/gal}$$

Last, substitute known parameters to solve for percent mixture strength.

The equation is:

$$\textbf{Percent mixture strength} = \frac{\textbf{[Soln. 1, gal)(lb/gal)(Avail. \%/100\%) + (Soln. 2, gal)(lb/gal)(Avail. \%/100\%)](100\% strength)}}{\textbf{(Soln. 1, gal)(lb/gal) + (Soln. 2, gal)(lb/gal)}}$$

where available chlorine % = Cl_2%.

$$\text{Percent mixture strength} = \frac{[685 \text{ gal}(11.34 \text{ lb/gal})(62.5\%/100\%) + 1{,}488 \text{ gal}(10.76 \text{ lb/gal})(78\%/100\%)](100\% \text{ strength})}{685 \text{ gal}(11.34 \text{ lb/gal}) + 1{,}488 \text{ gal}(10.76 \text{ lb/gal})}$$

$$\text{Percent mixture strength} = \frac{(4{,}854.94 \text{ lb} + 12{,}488.49 \text{ lb})(100\%)}{7{,}767.9 \text{ lb} + 16{,}010.88 \text{ lb}} = \frac{(17{,}343.43 \text{ lb})(100\%)}{23{,}778.78 \text{ lb}}$$

Percent mixture strength = 72.94%, round to **73% strength of new ZOP solution**

PUMPING DISCHARGE PROBLEMS

76. The level in a storage tank drops 5.45 ft in exactly 10.0 hr. If the tank has a diameter of 90.0 ft and the plant is producing 7.85 mgd, what is the average discharge rate of the two treated water discharge pumps in gallons per minute?

First, find the water production during the 10-hr interval.

Gallons of water treated in 10-hr interval = (7.85 mgd)(1,000,000)(10 hr)/24 hr

Gallons of water treated in 10-hr interval = 3,270,833 gal

Next, find the gallons contained in the 5.45 ft drop in water level.

Volume, tank = (0.785)(Diameter)^2(Height)

Volume of 5.45 ft in 90 ft diameter tank = (0.785)(90.0 ft)(90.0 ft)(5.45 ft)(7.48 gal/ft^3)

Volume of 5.45 ft in 90 ft diameter tank = 259,211 gal

Production plus the loss in level is the amount the discharge pump had to send into the distribution system.

Total gallons discharge pumps moved in 10 hr = 3,270,833 gal + 259,211 gal

= 3,530,044 gal

Next, find the number of minutes in 10 hr.

(10.0 hr)(60 min/hr) = 600 min

Last, divide the number of gallons the discharge pumps moved by the time in minutes.

Discharge pumps, gpm = 3,530,044 gal/600 min = 5,883 gpm, round to **5,880 gpm**

77. Find the average discharge rate of the treated water discharge pumps given the following data:

The clear well level rises 4.87 ft in 3.50 hr.

The clear well has a length of 338 ft and a width of 82.5 ft.

The plant produces 38.6 mgd.

First, find the water production during the 3.50-hr interval.

Gallons of water treated in 3.50-hr interval = (38.6 mgd)(1,000,000)(3.50 hr)/24 hr

Gallons of water treated in 3.50-hr interval = 5,629,167 gal

Next, find the gallons contained in the clear well's rise of 4.87 ft.

Volume of clear well = (Length)(Width)(Height of increase)

Volume of clear well = (338 ft)(82.5 ft)(4.87 ft)(7.48 gal/ft^3) = 1,015,784 gal

Production minus the rise in the water level is the amount the discharge pumps had to send into the distribution system, but first find the number of minutes in 3.50 hr.

(3.50 hr)(60 min/hr) = 210 min

Then, determine total gallons discharge pumps moved.

Total gallons discharge pumps moved in 3.50 hr = 5,629,167 gal – 1,015,784 gal

= 4,613,383 gal

Last, divide the number of gallons the discharge pumps moved by the time in minutes.

Discharge pumps, gpm = 4,613,383 gal/210 min = 21,968 gpm, round to **22,000 gpm**

78. A sodium hypochlorite solution is being pumped from a small tank that is 4.5 ft in diameter. If the level in the tank drops 10.3 in. in 12 hr, how many gallons per minute of hypochlorite solution was used?

First, convert the drop in tank level given in inches to feet.

Number of ft = (10.3 in.)(1 ft/12 in.) = 0.858 ft

Next, determine the number of gallons that was pumped.

The equation is:

Number of gal = (0.785)(Diameter, ft)^2(Level drop, ft)(7.48 gal/ft^3)

Number of gal = (0.785)(4.5 ft)(4.5 ft)(0.858 ft)(7.48 gal/ft^3) = 102 gal

Next, find the number of minutes in 12 hr.

$$\text{Number of min} = (12 \text{ hr})(60 \text{ min/hr}) = 720 \text{ min}$$

Last, divide the number of gallons pumped by the time in minutes.

$$\text{Number of gpm} = \frac{102 \text{ gal}}{720 \text{ min}} = \mathbf{0.14 \text{ gpm of NaOCl}}$$

PUMPING CALCULATIONS

79. What is the motor horsepower if 145 hp is required to run a pump with a motor efficiency of 86% and a pump efficiency of 77%?

Note: The horsepower = water horsepower (whp).

The equation is:

$$\textbf{Motor horsepower} = \frac{\textbf{(whp)}}{\textbf{(Motor efficiency)(Pump efficiency)}}$$

First, change percent motor and pump efficiencies to decimal form by dividing by 100%.

$$\text{Motor efficiency} = 86\%/100\% = 0.86$$

$$\text{Pump efficiency} = 77\%/100\% = 0.77$$

$$\text{Motor horsepower} =$$

$$\frac{(145 \text{ whp})}{(0.86 \text{ Motor efficiency})(0.77 \text{ Pump efficiency})} = 218.97 \text{ mhp, round to } \mathbf{220 \text{ mhp}}$$

80. What is the brake horsepower if 120 hp is supplied to a motor with 87% efficiency?

The equation is:

$$\textbf{Brake horsepower} = \textbf{(hp)(Motor efficiency)}$$

$$\text{Brake horsepower} = (120 \text{ hp})(87\%/100\% \text{ Motor efficiency}) = \mathbf{104 \text{ bhp}}$$

81. What is the motor horsepower if the brake horsepower is 76 and the motor efficiency is 85%?

The equation is:

Motor horsepower = Brake horsepower/Motor efficiency

Motor efficiency = 85%/100% = 0.85

Motor horsepower = 76 bhp/0.85 Motor efficiency = 89.41 mhp, round to **89 mhp**

82. Determine the motor horsepower for a pump station given the following parameters:

Motor efficiency:	**90.0%**	**Total head:**	**95 ft**
Pump efficiency:	**79.0%**	**Flow:**	**6.48 mgd**

First, convert million gallons per day to gallons per minute.

Gallons per minute = (6.48 mgd)(1,000,000/1 M)(1 d/1,440 min) = 4,500 gpm

Next, convert the percent motor and pump efficiencies to decimal form by dividing by 100%.

Motor efficiency = 90.0%/100% = 0.900

Pump efficiency = 79.0%/100% = 0.790

The equation for determining the motor horsepower with the given data is different from the previous problem.

The equation is:

$$\textbf{Motor horsepower} = \frac{\textbf{(Flow, gpm)(Total head, ft)}}{\textbf{(3,960)(Motor efficiency)(Pump efficiency)}}$$

Note: The number 3,960 is a constant.

$$\text{Motor horsepower} = \frac{(4{,}500\text{ gpm})(95\text{ ft})}{(3{,}960)(0.900\text{ Motor efficiency})(0.790\text{ Pump efficiency})}$$

= 151.8 mhp, round to **150 mhp**

83. Water that is being pumped from a water source has an elevation of 3,840 ft to an elevation of 4,275 ft. What is the total head if friction and minor head losses are 29 ft?

The equation is:

Total head, ft = Total static head, ft (difference in elevation) + Head losses, ft

Total head, ft = (4,275 ft – 3,840 ft) + 29 ft = 435 + 29 ft = **464 ft**

84. Find the water horsepower for the following system if the motor efficiency is 91%, the pump efficiency is 82%, and the motor horsepower is 120.

The equation is:

Water horsepower = (mhp)(Motor efficiency)(Pump efficiency)

First, convert the motor and pump efficiencies to decimal form by dividing by 100%.

Motor efficiency = 91%/100% = 0.91

Pump efficiency = 82%/100% = 0.82

Water horsepower = (120 mhp)(0.91 Motor efficiency)(0.82 Pump efficiency)

= 89.5 whp, round to **90 whp**

85. What is the brake horsepower if 350 hp is supplied to a motor with 85% efficiency?

The equation is:

Brake horsepower = (hp)(Motor efficiency)

Brake horsepower = (350 hp)(85%/100% Motor efficiency)

= 297.5 bhp, round to **300 bhp**

86. What is the motor horsepower if 240 hp is required to run a pump with a motor efficiency of 90% and a pump efficiency of 81%?

Note: The horsepower = water horsepower (whp).

The equation is:

$$\textbf{Motor horsepower} = \frac{\textbf{(whp)}}{\textbf{(Motor efficiency)(Pump efficiency)}}$$

First, change percent motor and pump efficiencies to decimal form by dividing by 100%.

$$\text{Motor efficiency} = 90\%/100\% = 0.90$$

$$\text{Pump efficiency} = 81\%/100\% = 0.81$$

$$\text{Motor horsepower} =$$

$$\frac{(240 \text{ whp})}{(0.90 \text{ Motor efficiency})(0.81 \text{ Pump efficiency})} = 329 \text{ mhp, round to } \mathbf{330\ mhp}$$

87. Find the water horsepower for a pumping system if the motor efficiency is 88%, the pump efficiency is 79%, and the motor horsepower is 280.

The equation is:

$$\textbf{Water horsepower} = \textbf{(mhp)(Motor efficiency)(Pump efficiency)}$$

First, convert the percent motor and pump efficiencies to decimal form by dividing by 100%.

$$\text{Motor efficiency} = 88\%/100\% = 0.88$$

$$\text{Pump efficiency} = 79\%/100\% = 0.79$$

$$\text{Water horsepower} = (280 \text{ mhp})(0.88 \text{ Motor efficiency})(0.79 \text{ Pump efficiency})$$

$$= 194.66 \text{ whp, round to } \mathbf{195\ whp}$$

88. Determine the motor horsepower for a pump station given the following parameters:

Motor efficiency:	**86%**	**Total head:**	**106 ft**
Pump efficiency:	**76%**	**Flow:**	**4.55 mgd**

First, convert million gallons per day to gallons per minute.

Gallons per minute = (4.55 mgd)(1,000,000/1 M)(1 d/1,440 min) = 3,159.7 gpm

Next, convert the percent motor and pump efficiencies to decimal form by dividing by 100%

Motor efficiency = 86%/100% = 0.86

Pump efficiency = 76%/100% = 0.76

The equation is:

$$\textbf{Motor horsepower} = \frac{\textbf{(Flow, gpm)(Total head, ft)}}{\textbf{(3,960)(Motor efficiency)(Pump efficiency)}}$$

Note: The number 3,960 is a constant.

$$\text{Motor horsepower} = \frac{(3{,}159.7 \text{ gpm})(106 \text{ ft})}{(3{,}960)(0.86 \text{ Motor efficiency})(0.76 \text{ Pump efficiency})} = 129.4 \text{ mhp, round to } \mathbf{130\ mhp}$$

89. Determine the horsepower required for a clear well water pump that needs to pump water to a storage tank given the following parameters:

Elevation of clear well water pump = 504.29 ft

Elevation of water storage tank = 674.51 ft

Length of pipeline from clear well water pump to storage tank = 6,043 ft

Pump above clear well (suction lift) = 6.4 ft

Friction loss in pipeline = 1.34 ft per 1,000 ft

Assume velocity head = 2.57 ft

Required flow per day (maximum) = 5,000 gpm

Pump efficiency = 86%

Motor efficiency = 92%

First, determine the pumping head for this pump.

$$\text{Pumping head} = \text{Elevation of water plant} - \text{Elevation of pump}$$

$$\text{Pumping head} = 674.51 \text{ ft} - 504.29 \text{ ft} = 170.22 \text{ ft}$$

Next, calculate the friction loss in the pipeline.

$$\text{Friction loss} = \frac{(6{,}043 \text{ ft})(1.34 \text{ ft})}{1{,}000 \text{ ft}} = 8.10 \text{ ft}$$

Now, calculate the total head.

$$\text{Total dynamic head} = \text{Suction lift} + \text{Pumping head} + \text{Friction loss} + \text{Velocity head}$$

$$\text{Total dynamic head} = 6.4 \text{ ft} + 170.22 \text{ ft} + 8.10 \text{ ft} + 2.57 \text{ ft}$$

$$\text{Total dynamic head} = 187.29 \text{ ft}$$

Last, determine the required horsepower of the pump.

$$\text{Horsepower} = \frac{(\text{gpm})(\text{Total dynamic head})}{(3{,}960)(\text{Pump efficiency})(\text{Motor efficiency})}$$

$$\text{Horsepower} = \frac{(5{,}000 \text{ gpm})(187.29 \text{ ft})}{(3{,}960)(86\%/100\%)(92\%/100\%)}$$

Horsepower = 298.9 (use a 300 hp motor)

90. Determine the horsepower needed for a raw water pump given the following parameters:

Elevation of raw water pump = 194.3 ft

Elevation of water treatment plant = 438.9 ft

Length of pipeline from raw water pump to treatment plant = 4,718 ft

Pump above the source water (suction lift) = 4 ft

Friction loss in pipeline = 1.31 ft per 1,000 ft

Assume velocity head = 2.44 ft

Required flow per day (maximum) = 6,000 gpm

Pump efficiency = 84%

Motor efficiency = 93%

First, determine the pumping head for this pump.

$$\text{Pumping head} = \text{Elevation of water plant} - \text{Elevation of pump}$$

$$\text{Pumping head} = 438.9 \text{ ft} - 194.3 \text{ ft} = 244.6 \text{ ft}$$

Next, calculate the friction loss in the pipeline.

$$\text{Friction loss} = \frac{(4{,}718 \text{ ft})(1.31 \text{ ft})}{1{,}000 \text{ ft}} = 6.18 \text{ ft}$$

Now, calculate the total head.

$$\text{Total dynamic head} = \text{Suction lift} + \text{Pumping head} + \text{Friction loss} + \text{Velocity head}$$

$$\text{Total dynamic head} = 4 \text{ ft} + 244.6 \text{ ft} + 6.18 \text{ ft} + 2.44 \text{ ft}$$

$$\text{Total dynamic head} = 257.22 \text{ ft}$$

Last, determine the required horsepower of the pump.

$$\text{Horsepower} = \frac{(\text{gpm})(\text{Total dynamic head})}{(3{,}960)(\text{Pump efficiency})(\text{Motor efficiency})}$$

$$\text{Horsepower} = \frac{(6{,}000 \text{ gpm})(257.22 \text{ ft})}{(3{,}960)(84\%/100\%)(93\%/100\%)}$$

$$\textbf{Horsepower = 499 (use a 500 hp motor)}$$

NET POSITIVE SUCTION HEAD CALCULATIONS

91. What is the net positive suction head available (NPSHA) given the following data? Will the pump cavitate if the net positive suction head required (NPSHR) is 20.3 ft?

Atmospheric pressure (AP) = 29.6 in. Hg

Static suction lift (SSL) = 11.5 ft

Friction head loss (H_f) = 0.55 ft

Vapor pressure at 9°F (VP) = 0.49 ft

First, determine the atmospheric pressure in feet.

The following applies:

1.13 ft/in. Hg

Thus,

Atmospheric pressure = (29.6 in. Hg)(1.13 ft/in. Hg) = 33.448 ft

Next, determine the net positive suction head required.

$$\text{NPSHA} = \text{AP, ft} - \text{SSL, ft} - H_f\text{, ft} - \text{VP, ft}$$

$$\text{NPSHA} = 33.448 \text{ ft} - 11.5 \text{ ft} - 0.55 \text{ ft} - 0.49 \text{ ft}$$

$$\text{NPSHA} = \mathbf{20.9 \text{ ft}}$$

Therefore:

NPSHA > NPSHR, so no cavitation should occur.

92. What is the net positive suction head available given the following data? Will the pump cavitate if the net positive suction head required is 23.6 ft?

Atmospheric pressure (AP) = 29.8 in. Hg

Static suction lift (SSL) = 12.4 ft

Friction head loss (H_f) = 0.63 ft

Vapor pressure at 12°F (VP) = 0.53 ft

First, determine the atmospheric pressure in feet.

The following applies:

$$1.13 \text{ ft/in. Hg}$$

Thus,

$$\text{Atmospheric pressure} = (29.8 \text{ in. Hg})(1.13 \text{ ft/in. Hg}) = 33.674 \text{ ft}$$

Next,determine the net positive suction head required.

$$\text{NPSHA} = \text{AP, ft} - \text{SSL, ft} - H_f\text{, ft} - \text{VP, ft}$$

$$\text{NPSHA} = 33.674 \text{ ft} - 12.4 \text{ ft} - 0.63 \text{ ft} - 0.53 \text{ ft}$$

$$\text{NPSHA} = \mathbf{20.1 \text{ ft}}$$

Therefore:

NPSHA < NPSHR, so cavitation should occur.

ESTIMATING WATER FLOW FROM A PIPE

93. Estimate the water flow from a fire hydrant based on the following data:

Pipe diameter = 2.0 in.

Height of pipe above ground = 28 in.

Distance water shoots out of pipe (length) = 28 in.

The equation is:

$$\textbf{Flow, gpm} = \frac{\textbf{2.83(Diameter)}^2\textbf{(Length, in.)}}{\textbf{(Height, in.)}^{1/2}}$$

Note: (Height)$^{1/2}$ means the square root of the height. The number 2.83 is a constant.

(continued)

Substitute known values and solve.

$$\text{Flow, gpm} = \frac{2.83(2.0\text{ in.})(2.0\text{ in.})(28\text{ in.})}{(28\text{ in.})^{1/2}} = \frac{316.96}{5.29} = 59.9\text{ gpm, round to }\mathbf{60\ gpm}$$

94. Estimate the flow from a fire hydrant in gallons per minute given the following parameters:

The hydrant opening is 3.0 in. The center of the opening is 24 in. vertically above the point where the water contacts the ground, and the water contacts the ground 42 in. horizontally from the hydrant's opening.

The equation is:

$$\mathbf{Flow,\ gpm} = \frac{\mathbf{2.83(Diameter)^2(Distance,\ in.)}}{\mathbf{(Height,\ in.)^{1/2}}}$$

$$\text{Flow, gpm} = \frac{2.83(3.0\text{-in.})^2(42\text{-in.})}{(24\text{-in.})^{1/2}} = \frac{1{,}069.74}{(24\text{-in.})^{1/2}} = \frac{1{,}069.74}{4.899}$$

$$= 218.36\text{ gpm, round to }\mathbf{220\ gpm}$$

95. A 1.0-in. reducer was put on the fire hydrant in problem 94 without changing the flow. How far will the water go out horizontally in feet?

First, rearrange the equation to solve for the horizontal distance in inches.

$$\mathbf{Distance,\ in.} = \frac{\mathbf{(Flow,\ gpm)(Height)^{1/2}}}{\mathbf{2.83(Diameter)^2}}$$

$$\text{Distance, in.} = \frac{(220\text{ gpm})(24)^{1/2}}{2.83(\text{Diameter})^2} = \frac{220\text{ gpm}(4.899)}{2.83(1.0)^2} = 380.84\text{ in.}$$

Last, convert inches to feet.

$$\frac{380.84\text{ in.}}{12\text{ in./1 ft}} = 31.7\text{ ft, round to }\mathbf{32\ ft}$$

96. An operator is filling a tank that is predominantly below ground with water. The tank is 28 ft in diameter and is 14.5 ft to the overflow. Estimate the time it will take to fill the tank in hours given the following data:

Pipe diameter = 1.75 in.

Height of pipe above tank = 35 in.

Distance water shoots out of pipe to top of tank (length) = 30.5 in.

First, determine the capacity of the tank in gallons to the overflow.

The equation is:

$$\textbf{Number of gal} = \mathbf{(0.785)(Diameter)^2(Height\ to\ overflow,\ ft)(7.48\ gal/ft^3)}$$

$$\text{Number of gal} = (0.785)(28\text{ ft})(28\text{ ft})(14.5\text{ ft})(7.48\text{ gal/ft}^3) = 66{,}751\text{ gal}$$

Next, determine the flow into the tank.

The equation is:

$$\textbf{Flow, gpm} = \frac{\mathbf{2.83(Diameter)^2(Length,\ in.)}}{\mathbf{(Height,\ in.)^{1/2}}}$$

Substitute known values and solve.

$$\text{Flow, gpm} = \frac{2.83(1.75\text{ in.})(1.75\text{ in.})(30.5\text{ in.})}{(35\text{ in.})^{1/2}} = \frac{264.34}{5.916} = 44.68\text{ gpm}$$

Last, find the number of hours it will take to fill the tank.

$$\text{Number of hr} = \frac{66{,}751\text{ gal}}{(44.68\text{ gpm})(60\text{ min/hr})} = 24.9\text{ hr, round to }\mathbf{25\ hr}$$

97. An operator is filling a small tank with a garden hose. The tank is 6.0 ft in diameter and is 4.8 ft high. Estimate the time it will take to fill the tank in minutes if the tank is filled to 95% capacity given the following data:

Hose diameter = 0.50 in.

Height of hose above tank = 14 in.

Distance water shoots out of hose to top of tank (length) = 43 in.

First, determine the capacity of the tank in gallons to the overflow.

The equation is:

$$\textbf{Number of gal} = \textbf{(0.785)(Diameter)}^2\textbf{(Height to overflow, ft)(7.48 gal/ft}^3\textbf{)}$$

$$\text{Number of gal} = (0.785)(6.0 \text{ ft})(6.0 \text{ ft})(4.8 \text{ ft})(7.48 \text{ gal/ft}^3) = 1{,}015 \text{ gal}$$

Next, find the number of gallons, if the tank is only filled to 95%.

$$\text{Number of gal} = (1{,}015 \text{ gal})(95\%/100\%) = 964.25 \text{ gal}$$

Next, determine the flow into the tank.

The equation is:

$$\textbf{Flow, gpm} = \frac{\textbf{2.83(Diameter)}^2\textbf{(Length, in.)}}{\textbf{(Height, in.)}^{1/2}}$$

Substitute known values and solve.

$$\text{Flow, gpm} = \frac{2.83(0.50 \text{ in.})(0.50 \text{ in.})(43 \text{ in.})}{(14 \text{ in.})^{1/2}} = \frac{30.4225}{3.742} = 8.13 \text{ gpm}$$

Last, find the number of minutes it will take to fill the tank.

$$\text{Number of hr} = \frac{964.25 \text{ gal}}{8.13 \text{ gpm}} = 118.6 \text{ min, round to } \textbf{120 min}$$

CALCULATING THE 90TH PERCENTILE FOR LEAD AND COPPER

The action level for lead is 0.015 mg/L and 1.3 mg/L for copper.

98. Given the following data for lead concentrations, find the 90th percentile and determine if the utility has to proceed with a corrosion program:

0.023 mg/L	**0.018mg/L**	**0.004 mg/L**	**0.019 mg/L**	**0.013 mg/L**
0.006 mg/L	**0.007 mg/L**	**0.026 mg/L**	**0.010 mg/L**	**0.006 mg/L**

First, line up the results in one column from largest at the top to smallest at the bottom.

0.026 mg/L

0.023 mg/L 90th percentile

0.019 mg/L

0.018 mg/L

0.013 mg/L

0.010 mg/L

0.007 mg/L

0.006 mg/L

0.006 mg/L

0.004 mg/L

For convenience there are ten analytical results, each equivalent to 10%. Count up from the bottom nine places (9 times 10% = 90% or the 90th percentile). The 90th percentile is 0.023 mg/L.

Because the 90th percentile of 0.023 mg/L exceeds the action level of 0.015 mg/L, the utility will be required to install a corrosion control program.

HYDRAULIC PRESS CALCULATIONS

99. A force of 85 lb is applied to a small cylinder on a hydraulic jack. The diameter of the small cylinder is 8.0 in. If the diameter of the large cylinder is 1.8 ft, what is the total lifting force?

To find the pressure on the small cylinder, use the following equation.

The equation is:

$$\textbf{Pressure} = \frac{\textbf{Force, lb}}{\textbf{Area, ft}^2}$$

First, convert 8.0 in. to feet.

$$(8.0 \text{ in.})(1 \text{ ft}/12 \text{ in.}) = 0.667 \text{ ft}$$

Substituting:

$$\text{Pressure} = \frac{85 \text{ lb}}{(0.785)(0.667 \text{ ft})(0.667 \text{ ft})} = 243.39 \text{ lb/ft}^2$$

Next, calculate the total force on the large cylinder.

The equation is:

$$\textbf{Total force} = \textbf{(Pressure)(Area, ft}^2\textbf{)}$$

$$\text{Total force} = (243.39 \text{ lb/ft}^2)(0.785)(1.8 \text{ ft})(1.8 \text{ ft}) = 619 \text{ lb, round to } \mathbf{620\ lb}$$

100. A force of 125 lb is applied to a small cylinder on a hydraulic jack. The diameter of the small cylinder is 10.0 in. If the diameter of the large cylinder is 2.25 ft, what is the total lifting force?

To find the pressure on the small cylinder, use the following equation.

The equation is:

$$\textbf{Pressure} = \frac{\textbf{Force, lb}}{\textbf{Area, ft}^2}$$

First, convert 10.0 in. to feet.

$$(10.0 \text{ in.})(1 \text{ ft}/12 \text{ in.}) = 0.833 \text{ ft}$$

$$\text{Pressure} = \frac{125 \text{ lb}}{(0.785)(0.833 \text{ ft})(0.833 \text{ ft})} = 229.48 \text{ lb/ft}^2$$

Next, calculate the total force on the large cylinder.

The equation is:

$$\textbf{Total force} = \textbf{(Pressure, lb/ft}^2\textbf{)(Area, ft}^2\textbf{)}$$

Total force = (229.48 lb/ft²)(0.785)(2.25 ft)(2.25 ft) = 911.97 lb, round to **912 lb**

101. Determine the diameter in inches of a small cylinder hydraulic jack given the following data:

Force exerted on small cylinder = 314 lb

Large cylinder = 2.50 ft

Total force = 2,845 lb

We need to work backward from the previous two examples, so first find the pressure.

$$\textbf{Pressure, lb/ft}^2 = \frac{\textbf{Total force}}{\textbf{(0.785)(Diameter, large cylinder)}^2}$$

$$\text{Pressure, lb/ft}^2 = \frac{2{,}845 \text{ lb}}{(0.785)(2.50 \text{ ft})(2.50 \text{ ft})} = 579.87 \text{ lb/ft}^2$$

Now, find the area of the small cylinder.

$$\text{Pressure, lb/ft}^2 = \frac{\text{Force, lb}}{(0.785)(\text{Diameter, ft})^2}$$

Rearrange the equation to solve for area in square feet.

$$(\text{Diameter, ft})^2 = \frac{\text{Force, lb}}{\text{Pressure, lb/ft}^2}$$

$$(\text{Diameter, ft})^2 = \frac{314 \text{ lb}}{(0.785)(579.87 \text{ lb/ft}^2)} = 0.69 \text{ ft}^2$$

Last, take the square root of 0.69 ft².

$$\text{Diameter, ft} = (0.69 \text{ ft}^2)^{1/2} = 0.83 \text{ ft}$$

Number of in. = (12 in./ft)(0.83 ft) = 9.96 in., round to **10 in.**

KILOWATT DETERMINATIONS

102. Calculate the total kilowatts needed to operate the following small facility if everything was running:

Raw water pump	200 hp
Four flocculators, 10 hp each	40 hp
Filter pump for backwashing	60 hp
Chlorination	20 hp
Clear water pump	120 hp
Lighting	16 hp
Instrumentation	5 hp
Total horsepower	461 hp

The equation is:

$$\text{Kilowatts} = (\text{Number of hp})(0.746\ \text{kW/hp})$$

$$\text{Kilowatts} = (461\ \text{hp})(0.746\ \text{kW/hp}) = \mathbf{344\ kW}$$

103. How many kilowatts will it take to operate a 100-hp pump, assuming the start up energy is 2 times?

$$\text{Kilowatts} = (\text{Number of hp})(0.746\ \text{kW/hp})(2)$$

$$\text{Kilowatts} = (100\ \text{hp})(0.746\ \text{kW/hp})(2) = 149.2\ \text{kW, round to } \mathbf{150\ kW}$$

HAZEN–WILLIAMS DETERMINATIONS

In order to calculate Hazen–Williams problems you will have to have a calculator that can calculate exponents.

104. Determine the approximate C factor given the following data:

Diameter of pipe	= **18 in. (1.5 ft)**		**Upstream pressure gauge**	=	**87 ft**
Flow	= **2,250 gpm**		**Downstream pressure gauge**	=	**84 ft**
			Distance between gauges	=	**1,500 ft**

First, determine the energy loss in feet.

The equation is:

Energy loss, ft = Upstream gauge read – Downstream gauge read

Energy loss, ft = 87 ft – 84 ft = 3 ft

Next, calculate the slope.

The equation is:

$$\textbf{Slope} = \frac{\textbf{Energy loss, ft}}{\textbf{Distance, ft}}$$

$$\text{Slope} = \frac{3 \text{ ft}}{1{,}500 \text{ ft}} = 0.002 \text{ ft}$$

Last, find the C factor.

The equation is:

$$\textbf{C factor} = \frac{\textbf{Flow, gpm}}{\textbf{(193.75)(Diameter of pipe, ft)}^{2.63}\textbf{(Slope, ft)}^{0.54}}$$

where 193.75 is a constant, and 2.63 and 0.54 are exponents.

$$\text{C factor} = \frac{2{,}250 \text{ gpm}}{(193.75)(1.5 \text{ ft})^{2.63}(0.002, \text{ ft})^{0.54}}$$

$$\text{C factor} = \frac{2{,}250 \text{ gpm}}{(193.75)(2.90)(0.0349)} = 114.74, \text{ round to } \mathbf{110}$$

105. Determine the approximate C factor for a pipe that is 2.0 ft in diameter and has a flow of 3,425 gpm given the following data:

Upstream pressure gauge	=	**154 ft**
Downstream pressure gauge	=	**149 ft**
Distance between gauges	=	**3,820 ft**

First, determine the energy loss in feet.

$$\text{Energy loss, ft} = \text{Upstream gauge read} - \text{Downstream gauge read}$$

$$\text{Energy loss, ft} = 154 \text{ ft} - 149 \text{ ft} = 5 \text{ ft}$$

Next, calculate the slope.

The equation is:

$$\textbf{Slope} = \frac{\textbf{Energy loss, ft}}{\textbf{Distance, ft}}$$

$$\text{Slope} = \frac{5 \text{ ft}}{3{,}820 \text{ ft}} = 0.00131 \text{ ft}$$

Last, find the C factor.

The equation is:

$$\textbf{C factor} = \frac{\textbf{Flow, gpm}}{\textbf{(193.75)(Diameter of pipe, ft)}^{2.63}\textbf{(Slope, ft)}^{0.54}}$$

$$\text{C factor} = \frac{3{,}425 \text{ gpm}}{(193.75)(2.0 \text{ ft})^{2.63}(0.00131 \text{ ft})^{0.54}}$$

$$\text{C factor} = \frac{3{,}425 \text{ gpm}}{(193.75)(6.19)(0.0278)} = 102.7, \text{ round to } \mathbf{103}$$

106. Determine the approximate C factor given the following data:

Diameter of pipe	**= 45.72 cm**	**Upstream pressure gauge**	**= 45 m**
Flow	**= 7,500 L**	**Downstream pressure gauge**	**= 42.5 m**
		Distance between gauges	**= 1,682 m**

First, convert centimeters to meters.

$$\text{Number of m} = (45.72 \text{ cm})(1 \text{ m}/100 \text{ cm}) = 0.4572 \text{ m}$$

Next, determine the energy loss, in meters.

$$\textbf{Energy loss, m} = \textbf{Upstream gauge read} - \textbf{Downstream gauge read}$$

$$\text{Energy loss, m} = 45 \text{ m} - 42.5 \text{ m} = 2.5 \text{ m}$$

Next, calculate the slope.

The equation is:

$$\textbf{Slope} = \frac{\textbf{Energy loss, m}}{\textbf{Distance, m}}$$

$$\text{Slope} = \frac{2.5 \text{ m}}{1{,}682 \text{ m}} = 0.00149 \text{ m}$$

Then, convert liters per minute to million liters per day.

$$\text{Number of ML/d} = \frac{(7{,}500 \text{ L/min})(1{,}440 \text{ min/d})}{1{,}000{,}000/1 \text{ M}} = 10.8 \text{ ML/d}$$

Last, find the C factor.

The equation is:

$$\textbf{C factor} = \frac{\textbf{Flow, ML/d}}{\textbf{(23.97)(Diameter of pipe, m)}^{2.63}\textbf{(Slope, m)}^{0.54}}$$

where 23.97 is a constant used for the above metric units.

$$\text{C factor} = \frac{10.8 \text{ ML/d}}{(23.97)(0.4572 \text{ m})^{2.63}(0.00149 \text{ m})^{0.54}}$$

$$\text{C factor} = \frac{10.8 \text{ ML/d}}{(23.97)(0.128)(0.0298)} = 118.12, \text{ round to } \mathbf{120}$$

107. Determine the approximate C factor given the following data:

Diameter of pipe	= **250.0 mm**	**Upstream pressure gauge**	=	**78 m**
Flow	= **6,800 L**	**Downstream pressure gauge**	=	**56.5 m**
		Distance between gauges	=	**818 m**

First, convert millimeters to meters.

$$\text{Number of m} = (250.0 \text{ mm})(1 \text{ m}/1000 \text{ mm}) = 0.2500 \text{ m}$$

Next, determine the energy loss in meters.

$$\textbf{Energy loss, m} = \textbf{Upstream gauge read} - \textbf{Downstream gauge read}$$

$$\text{Energy loss, m} = 78 \text{ m} - 56.5 \text{ m} = 21.5 \text{ m}$$

Next, calculate the slope.

The equation is:

$$\textbf{Slope} = \frac{\textbf{Energy loss, m}}{\textbf{Distance, m}}$$

$$\text{Slope} = \frac{21.5 \text{ m}}{818 \text{ m}} = 0.0263 \text{ m}$$

Then, convert liters per minute to million liters per day.

$$\text{Number of ML/d} = \frac{(6{,}800 \text{ L/min})(1{,}440 \text{ min/d})}{1{,}000{,}000/1 \text{ M}} = 9.792 \text{ ML/d}$$

Last, find the C factor.

The equation is:

$$\textbf{C factor} = \frac{\textbf{Flow, ML/d}}{\textbf{(23.97)(Diameter of pipe, m)}^{2.63}\textbf{(Slope, m)}^{0.54}}$$

where 23.97 is a constant used for the above units.

$$\text{C factor} = \frac{9.792 \text{ ML/d}}{(23.97)(0.2500 \text{ m})^{2.63}(0.026 \text{ m})^{0.54}}$$

$$\text{C factor} = \frac{9.792 \text{ ML/d}}{(23.97)(0.026)(0.139)} = 112.6, \text{ round to } \mathbf{110}$$

108. Estimate the C factor for the following system:

Water main diameter = 8 in.

Flow = 850 gpm

Pressure difference = 6 ft for gauges 400 ft apart

First, convert the diameter of the water main to feet.

$$\text{Number of ft} = \frac{8 \text{ in.}}{12 \text{ in./ft}} = 0.67 \text{ ft}$$

Next, calculate the slope of the system.

$$\textbf{Slope, ft/ft} = \frac{\textbf{Energy loss, ft}}{\textbf{Distance, ft}}$$

$$\text{Slope, ft/ft} = \frac{6 \text{ ft}}{400 \text{ ft}} = 0.015$$

Last, determine the C factor.

The equation is:

$$\text{C factor} = \frac{\text{Flow, gpm}}{193.75 \text{ (Diameter, ft)}^{2.63}\text{(Slope, ft/ft)}^{0.54}}$$

$$\text{C factor} = \frac{850 \text{ gpm}}{193.75 \text{ (0.67 ft)}^{2.63}\text{(0.015 ft/ft)}^{0.54}}$$

$$\text{C factor} = \frac{850 \text{ gpm}}{193.75 \text{ (0.349 ft)(0.1035 ft/ft)}}$$

$$\text{C factor} = \frac{850 \text{ gpm}}{6.999}$$

$$\text{C factor} = \mathbf{121}$$

CORROSION RATES BASED ON COUPONS

109. Determine the corrosion rate in mils penetration per year on a steel coupon given the following data:

Mild steel coupon area = 3.44 in.^{2}

Exposure time = 2,160 hr

Metal density = 7.873 g/cm^3

Original weight of the coupon = 10.9348 g

Final weight of the coupon = 10.8417 g

First, determine the weight loss of the coupon.

$$\textbf{Weight loss} = \textbf{Original weight} - \textbf{Final weight}$$

$$\text{Weight loss} = 10.9348 \text{ g} - 10.8417 \text{ g} = 0.0931 \text{ g}$$

Write the equation for determining the mils penetration per year.

$$\textbf{Mils penetration per year} = \frac{\textbf{(Weight loss)(534)(1,000)}}{\textbf{(Coupon area, in.}^2\textbf{)(Time, hr)(Metal density, g/cm}^3\textbf{)}}$$

where 534 is a constant factor.

$$\text{Mils penetration per year} = \frac{(0.0931 \text{ g})(534)(1{,}000)}{(3.44 \text{ in.}^2)(2{,}160 \text{ hr})(7.873 \text{ g/cm}^3)} = \textbf{0.850 mpy}$$

110. Determine the corrosion rate in mils penetration per year on a copper coupon given the following data:

Copper coupon area = 3.40 in.^{2}

Exposure time = 1,440 hr

Metal density = 8.94 g/cm^3

Original weight of the coupon = 12.3915 g

Final weight of the coupon = 12.1694 g

First, determine the weight loss of the coupon.

$$\textbf{Weight loss} = \textbf{Original weight} - \textbf{Final weight}$$

$$\text{Weight loss} = 12.3915 \text{ g} - 12.1694 \text{ g} = 0.2221 \text{ g}$$

Write the equation for determining the mils penetration per year.

$$\textbf{Mils penetration per year} = \frac{\textbf{(Weight loss)(534)(1,000)}}{\textbf{(Coupon area, in.}^2\textbf{)(Time, hr)(Metal density, g/cm}^3\textbf{)}}$$

$$\text{Mils penetration per year} = \frac{(0.2221\text{ g})(534)(1{,}000)}{(3.40\text{ in.}^2)(1{,}440\text{ hr})(8.94\text{ g/cm}^3)} = \textbf{2.71 mpy}$$

111. Determine the corrosion rate in mils penetration per year on a lead coupon given the following data:

Lead coupon area = 3.42 in.^{2}

Exposure time = 2,160 hr

Metal density = 11.33 g/cm^3

Original weight of the coupon = 15.6877 g

Final weight of the coupon = 15.3558 g

First, determine the weight loss of the coupon.

$$\textbf{Weight loss} = \textbf{Original weight} - \textbf{Final weight}$$

$$\text{Weight loss} = 15.6877\text{ g} - 15.3558\text{ g} = 0.3319\text{ g}$$

Write the equation for determining the mils penetration per year.

$$\textbf{Mils penetration per year} = \frac{\textbf{(Weight loss)(534)(1,000)}}{\textbf{(Coupon area, in.}^2\textbf{)(Time, hr)(Metal density, g/cm}^3\textbf{)}}$$

$$\text{Mils penetration per year} = \frac{(0.3319\text{ g})(534)(1{,}000)}{(3.42\text{ in.}^2)(2{,}160\text{ hr})(11.33\text{ g/cm}^3)} = \textbf{2.12 mpy}$$

MILLIAMP SIGNAL CALCULATIONS

112. The SCADA system at a water plan uses a 4-mA to 20-mA signal to monitor the level of a backwash tank. If the readout on a SCADA board reads 8 mA, what is the height of the water in a tank if it has an overflow height of 25.5 feet? Give the answer to the nearest tenth of a foot.

The following applies:

$$4 \text{ mA} = 0 \text{ feet in the tank and that } 20 \text{ mA} = 25.5 \text{ feet in the tank}$$

First, subtract 4 mA from the board reading and from the maximum amount.

$$8 \text{ mA} - 4 \text{ V} = 4 \text{ mA}$$

$$20 \text{ mA} - 4 \text{ mA} = 16 \text{ mA}$$

Next, divide the current board reading by the maximum reading.

$$\frac{4 \text{ mA}}{16 \text{ mA}} = 0.25$$

Note: Each foot in the tank will cause a milliamp reading of 0.25.

The height of water in the tank is equal to this ratio multiplied by the maximum height of water in feet.

$$\text{Height, ft} = (0.25)(25.5 \text{ ft}) = \mathbf{6.4 \text{ ft}}$$

113. A pressure transducer has a scale of 0 to 200 psi. Currently there is 178 psi on the pumping system. What would the SCADA reading be on the board in milliamps for a 4-mA to 20-mA signal?

The equation is:

$$\frac{\text{mA maximum}}{\text{Maximum pressure, psi}} = \frac{\text{Unknown number of mA}}{\text{Existing pressure, psi}}$$

$$\text{Number of mA} = \frac{(\text{Pressure, psi})(\text{mA maximum})}{\text{Maximum pressure, psi}}$$

Substitute known values.

$$\text{Number of mA} = \frac{(178 \text{ psi})(20 \text{ mA})}{200 \text{ psi}} = \mathbf{17.8 \text{ mA}}$$

Distribution Grade 4 Test

(answers on p. 244)

1. How many pounds of lime must be added to exactly 220 gal of water to produce a 15.0% lime slurry?

2. Water is flowing at a velocity of 3.40 ft/s in an 8.0-in. diameter pipe. If the pipe changes from the 8.0-in. to a 10.0-in. pipe, what will the velocity be in the 10.0-in. pipe?

3. A well that is 319 ft in depth and 12 in. in diameter requires disinfection. Depth to the water from the top of the casing is 89 ft. If the desired dose is 40.0 mg/L, what is the number of ounces of calcium hypochlorite (58.0% available chlorine) required?

4. **Determine the volume of water in gallons for the following distribution system:**
   Distribution pipe A is 1,505 ft in length and 3.0 ft in diameter.
   Distribution pipe B is 2,460 ft in length and 2.0 ft in diameter.
   The storage tank is 90.0 ft in diameter and has a water height of 34.50 ft.

5. **A tank is 110 ft in diameter and is 42 ft tall. If there are 2,010,000 gal of water in the tank, what is the psi at the bottom of the tank? What is the psi 7 ft above the bottom of the tank?**

6. **Water is flowing at a velocity of 3.12 ft/s in a 6.0-in. diameter pipe. If the pipe changes from the 6.0-in. to a 10.0-in. pipe, what will the velocity be in the 10.0-in. pipe?**

7. **A pipe that is 2.50 ft in diameter and 2,245 ft long is to be disinfected with a 60.5% calcium hypochlorite solution. If the desired dose is 40.0 mg/L, what is the number of calcium hypochlorite gallons needed?**

8. How many calcium hypochlorite tablets will be needed to disinfect a water main given the following information. Assume the calcium hypochlorite solution weighs 8.34 lb/gal.

   Note: Round the partial tablet up to the next whole tablet.

   Length of pipe = 1,048 ft

   Pipe diameter = 3.0 ft

   Calcium hypochlorite = 60.0% available chlorine

   Dosage required = 40.0 mg/L

   Each tablet = 0.36 lb.

9. A solution containing exactly 725 gal of a 6.0% sodium hypochlorite solution is required. How many gallons of a 10.5% solution must be mixed with a 1.5% solution to make the required solution? Give the answer to the nearest tenth of a gallon.

10. What is the percent strength of a mixture of two zinc orthophosphate (ZOP) solutions given the following parameters?

    ZOP solution 1: 445 gal 65.4% solution and a specific gravity of 1.33

    ZOP solution 2: 1,947 gal of a 74% solution and a specific gravity of 1.27

Solutions to Distribution Grade 4 Test

1. The equation is:

$$\textbf{Percent lime} = \frac{\mathbf{(x \text{ lb})(100\%)}}{\mathbf{x \text{ lb} + (\text{Number of gal})(8.34 \text{ lb/gal})}}$$

$$15\% \text{ lime} = \frac{(x \text{ lb})(100\%)}{x \text{ lb} + (220 \text{ gal})(8.34 \text{ lb/gal})}$$

Multiply both sides of the equation by [x lb + (220 gal)(8.34 lb/gal)]. This leaves the following:

$$15.0\% \text{ lime}[x \text{ lb} + (220 \text{ gal})(8.34 \text{ lb/gal})] = (x \text{ lb})(100\%)$$

Do the multiplication.

$$15.0x \text{ \%, lb} + 27{,}522\% \text{ lb} = 100x \text{ \%, lb}$$

Subtract 15.0x %, lb from both sides of the equation.

$$27{,}522\%, \text{ lb} = 85.0x \text{ \%, lb}$$

Last, divide both sides of the equation by 85.0%.

x = 323.79 lb, round to **320 lb of lime**

2. Flow in the 8.0-in. pipe equals the flow in the 10.0-in. pipe, as the flow must remain constant.

$$\mathbf{Q_1 = Q_2}$$

Because Q, flow = (Area)(Velocity), it follows that:

(Area 1)(Velocity 1) = (Area 2)(Velocity 2)

First, find the diameters in feet for the 8.0-in. and 10.0-in. pipes.

Diameter for 8.0-in. pipe = 8.0 in.(1 ft/12-in.) = 0.667 ft

Diameter for 10.0-in. pipe = 10.0 in.(1 ft/12-in.) = 0.833 ft

Then, determine the areas of each pipe.

$$\text{Area} = (0.785)(\text{Diameter})^2$$

$$\text{Area 1 (8.0-in.)} = (0.785)(0.667 \text{ ft})(0.667 \text{ ft}) = 0.349 \text{ ft}^2$$

$$\text{Area 2 (10.0-in)} = (0.785)(0.833 \text{ ft})(0.833 \text{ ft}) = 0.545 \text{ ft}^2$$

Last, substitute the areas calculated and known velocity in the 8.0-in. pipe.

$$(0.349 \text{ ft}^2)(3.40 \text{ ft/s}) = (0.545 \text{ ft}^2)(x, \text{ ft/s})$$

Solve for x.

$$x, \text{ ft/s} = \frac{(0.349 \text{ ft}^2)(3.40 \text{ ft/s})}{(0.545 \text{ ft}^2)} = \textbf{2.2 ft/s in the 10.0-in. pipe}$$

3. First, find the length (in feet) of the water in the casing.

Length of water-filled casing = Depth of well – Depth of water to top of casing

Length of water-filled casing = 319 ft – 89 ft = 230 ft

Then, convert the diameter from inches to feet.

$$\text{Diameter, ft} = \frac{12 \text{ in.}}{12 \text{ in./ft}} = 1.0 \text{ ft}$$

Next, determine the volume in gallons of water in the well casing.

The equation is:

$$\textbf{Volume, gal} = \mathbf{(0.785)(Diameter)^2(Length)(7.48\ gal/ft^3)}$$

$$\text{Volume, gal} = (0.785)(1.0 \text{ ft})(1.0 \text{ ft})(230 \text{ ft})(7.48 \text{ gals/ft}^3) = 1{,}351 \text{ gal}$$

Next, determine the number of million gallons.

Million gallons = (1,351 gal)(1 M/1,000,000) = 0.00135 mil gal

Using the "pounds" equation, calculate the number of pounds of chlorine needed.

Pounds of chlorine = (0.00135 mil gal)(40.0 mg/L)(8.34 lb/gal) = 0.450 lb

(continued)

Last, calculate the ounces of calcium hypochlorite solution required.

$$\textbf{Number of gal, Ca(OCl)}_2 = \frac{\textbf{(Chlorine, lb)(100\%)(128 oz/gal)}}{\textbf{(8.34 lb/gal)(Hypochlorite, \%)}}$$

$$\text{Number of gal, Ca(OCl)}_2 = \frac{(0.450\text{ lb})(100\%)(128\text{ oz/gal})}{(8.34\text{ lb/gal})(58.0\%)} = 11.9\text{ oz, round to } \mathbf{12\ oz}$$

4. First, find the volume in gallons for each of the distribution pipes.
The equation is:

$$\textbf{Volume, gal} = \textbf{(0.785)(Diameter)}^2\textbf{(Length, ft)(7.48 gal/ft3)}$$

$$\text{Volume, pipe A, gal} = (0.785)(3.0\text{ ft})(3.0\text{ ft})(1{,}505\text{ ft})(7.48\text{ gal/ft}^3) = 79{,}534\text{ gal}$$

$$\text{Volume, pipe B, gal} = (0.785)(2.0\text{ ft})(2.0\text{ ft})(2{,}460\text{ ft})(7.48\text{ gal/ft}^3) = 57{,}779\text{ gal}$$

Next, determine the volume of the storage tank.

The equation is:

$$\textbf{Volume, gal} = \textbf{(0.785)(Diameter)}^2\textbf{(Height, ft)(7.48 gal/ft}^3\textbf{)}$$

$$\text{Volume, gal} = (0.785)(90.0\text{ ft})(90.0\text{ ft})(34.5\text{ ft})(7.48\text{ gal/ft}^3) = 1{,}640{,}874\text{ gal}$$

Last, add the three volumes together for the total volume.

$$\text{Total volume, gal} = 79{,}534\text{ gal} + 57{,}779\text{ gal} + 1{,}640{,}874\text{ gal}$$

$$\text{Total volume, gal} = 1{,}778{,}187\text{ gal, round to } \mathbf{1{,}800{,}000\ gal}$$

5. First, calculate the number of cubic feet of water present.

$$\frac{2{,}010{,}000\text{ gal}}{7.48\text{ gal/ft}^3} = 268{,}717\text{ ft}^3$$

Then, calculate the depth in cubic feet.

$$\textbf{Number of ft}^3 = \textbf{(0.785)(Diameter)}^2\textbf{(Depth)}$$

Solve for depth.

$$268{,}717\text{ ft}^3 = (0.785)(110\text{ ft})(110\text{ ft})(\text{Depth})$$

$$\text{Depth} = \frac{268{,}717\text{ ft}^3}{(0.785)(110\text{ ft})(110\text{ ft})}$$

$$\text{Depth} = 28.29\text{ ft}$$

Next, solve for the number of psi at the bottom of the tank.

$$\text{psi} = \frac{\text{Depth}}{2.31 \text{ ft/psi}} = \frac{28.29 \text{ ft}}{2.31 \text{ ft/psi}} = 12.2 \text{ psi, round to } \mathbf{12 \text{ psi}}$$

Last, solve for psi 7 ft above the bottom of the tank. The depth of the water is 7 ft less than the total depth (28.29 ft – 7 ft = 21.29 ft). Thus:

$$\text{psi} = \frac{\text{Depth}}{2.31 \text{ ft/psi}} = \frac{21.29 \text{ ft}}{2.31 \text{ ft/psi}} = \mathbf{9.2 \text{ psi}}$$

6. Flow in the 6.0-in. pipe equals the flow in the 10-in. pipe, as the flow must remain constant.

$$\mathbf{Q_1 = Q_2}$$

Because Q, flow = (Area)(Velocity), it follows that:

$$\mathbf{(Area\ 1)(Velocity\ 1) = (Area\ 2)(Velocity\ 2)}$$

First, find the diameters in feet for the 6-in. and 10.0-in. pipes.

$$\text{Diameter for 6.0-in. pipe} = 6.0\text{-in.}(1 \text{ ft/12-in.}) = 0.5 \text{ ft}$$

$$\text{Diameter for 10.0-in. pipe} = 10.0\text{-in.}(1 \text{ ft/12-in.}) = 0.833 \text{ ft}$$

Then, determine the areas of each pipe:

$$\mathbf{Area = (0.785)(Diameter)^2}$$

$$\text{Area 1 (6.0-in.)} = (0.785)(0.5 \text{ ft})(0.5 \text{ ft}) = 0.196 \text{ ft}^2$$

$$\text{Area 2 (10-in.)} = (0.785)(0.833 \text{ ft})(0.833 \text{ ft}) = 0.545 \text{ ft}^2$$

Last, substitute the areas calculated and known velocity in the 6.0-in. pipe.

$$(0.196 \text{ ft}^2)(3.12 \text{ ft/s}) = (0.545 \text{ ft}^2)(x, \text{ ft/s})$$

Solve for x.

$$x, \text{ ft/s} = \frac{(0.196 \text{ ft}^2)(3.12 \text{ ft/s})}{(0.545 \text{ ft}^2)} = 1.12 \text{ ft/s, round to } \mathbf{1.1 \text{ ft/s in the 10.0-in. pipe}}$$

7. First, calculate the number of gallons in the pipeline.

The equation is:

$$\textbf{Number of gal} = \mathbf{(0.785)(Diameter,\ ft)^2(Length,\ ft)(7.48\ gal/ft^3)}$$

$$\text{Number of gal} = (0.785)(2.50\text{ ft})(2.50\text{ ft})(2{,}245\text{ ft})(7.48\text{ gal/ft}^3) = 82{,}389\text{ gal}$$

Convert gallons to million gallons.

$$\text{Number of mil gal} = (82{,}389\text{ gal})(1\text{ M}/1{,}000{,}000) = 0.0824\text{ mil gal}$$

Next, determine the number of pounds of calcium hypochlorite needed using the "pounds" equation.

$$\text{Number of lb, } Ca(OCl)_2 = (\text{mil gal})(\text{Dosage, mg/L})(8.34\text{ lb/gal})$$

$$\text{Number of lb, } Ca(OCl)_2 = (0.0824\text{ mil gal})(40.0\text{ mg/L})(8.34\text{ lb/gal}) = 27.5\text{ lb of } Ca(OCl)_2$$

Last, determine the number of gallons needed.

$$\text{Number of gal, } Ca(OCl)_2 = \frac{[\text{Number of lb, } Ca(OCl)_2](100\%)}{(8.34\text{ lb/gal})(\%\text{ solution})}$$

$$\text{Number of gal, } Ca(OCl)_2 = \frac{[27.5\text{ lb } Ca(OCl)_2](100\%)}{(8.34\text{ lb/gal})(60.5\%)} = \mathbf{5.45\ gal\ of\ Ca(OCl)_2}$$

8. First, determine the volume of the pipeline in million gallons.

$$\text{Number of mil gal} = \frac{(0.785)(\text{Diameter})^2(\text{Length, ft})(7.48\text{ gal/ft}^3)}{1{,}000{,}000/1\text{ M}}$$

$$\text{Number of mil gal} = \frac{(0.785)(3.0\text{ ft})(3.0\text{ ft})(1{,}048\text{ ft})(7.48\text{ gal/ft}^3)}{1{,}000{,}000/1\text{ M}} = 0.0554\text{ mil gal}$$

Next, determine the number of pounds of calcium hypochlorite needed.

$$\textbf{Number of lb, } \mathbf{Ca(OCl)_2} = \frac{\mathbf{(mil\ gal)(Dosage,\ mg/L)(8.34\ lb/gal)}}{\textbf{Percent purity}}$$

$$\text{Number of lb, } Ca(OCl)_2 = \frac{(0.0554\text{ mil gal})(40.0\text{ mg/L})(8.34\text{ lb/gal})}{60.0\%/100\%} = 30.8\text{ lb}$$

Last, find the number of tablets needed.

$$\text{Number of tablets} = \frac{30.8\text{ lb}}{0.36\text{ lb per tablet}} = 85.6\text{ tablets, round up to } \mathbf{86\ tablets}$$

9. Solve the problem using the dilution triangle.

10.5%		4.5	4.5 parts of the 10.5% solution are required for every 9.0 parts
	6.0%		
1.5%		4.5	4.5 parts of the 1.5% solution are required for every 9.0 parts
		9.0 total parts	

$$\frac{4.5 \text{ parts}(725 \text{ gal})}{9.0 \text{ parts}} = \mathbf{362.5 \text{ gal of the } 10.5\% \text{ solution}}$$

$$\frac{4.5 \text{ parts}(725 \text{ gal})}{9.0 \text{ parts}} = \frac{\mathbf{362.5}}{725 \text{ gal}} \mathbf{\text{ gal of the } 1.5\% \text{ solution}}$$

To make the 725 gal of the 6.0% solution, mix 362.5 gal of the 10.5% solution with 362.5 gal of the 1.5% solution.

10. First, determine the weight in pounds per gallon for both of the solutions. The equation is:

$$\text{Weight of solution} = (\text{Specific gravity})(8.34 \text{ lb/gal})$$

$$\text{Weight of solution 1} = (1.33 \text{ sp gr})(8.34 \text{ lb/gal}) = 11.09 \text{ lb/gal}$$

$$\text{Weight of solution 2} = (1.27 \text{ sp gr})(8.34 \text{ lb/gal}) = 10.59 \text{ lb/gal}$$

The equation is:

$$\text{Percent mixture strength} = \frac{[\text{Soln. 1, gal(lb/gal)(Avail. \%/100\%)} + \text{Soln. 2, gal(lb/gal)(Avail. \%/100\%)}]}{\text{Soln. 1, gal(lb/gal)} + \text{Soln. 2, gal/(lb/gal)}} \times 100 \text{ strength}$$

Substituting:

$$\text{Percent mixture strength} = \frac{445 \text{ gal}(11.09 \text{ lb/gal})(65.4\%/100\%) + 1{,}947 \text{ gal}(10.59 \text{ lb/gal})(74\%/100\%)}{445 \text{ gal}(11.09 \text{ lb/gal}) + 1{,}947 \text{ gal}(10.59 \text{ lb/gal})} \times 100\%$$

$$\text{Percent mixture strength} = \frac{3{,}227.52 \text{ lb} + 15{,}257.86 \text{ lb}}{4{,}935.05 \text{ lb} + 20{,}618.73 \text{ lb}} \times 100\% = \frac{18{,}485.38 \text{ lb}}{25{,}553.78 \text{ lb}} \times 100\%$$

Percent mixture strength = 72.34%, round to **72% strength of new ZOP solution**

Appendixes

APPENDIX A

COMMON CONVERSION FACTORS

AREA

1 acre = 43,560 square feet (ft^2)
1 acre-ft = 43,560 cubic feet (ft^3)
2.4711 acre = 1 hectare (ha)
1 ha = 0.4047 acre
1 hectare = 10,000 square meters (m^2)
1 square mile (mi^2) = 640 acres

CONCENTRATION

1% solution = 1 part in 100 parts
1% solution = 10,000 ppm
1 ppm = 1 mg/L
1 gpg = 17.12 ppm

DENSITY

Water has a density of 1 g/mL or 8.34 lb/gal or 62.4 lb/ft^3

FLOW

1 miner's inch = 1.5 ft^3/min
1 ft^3/s = 448.8 gal/min (gpm)
1 ft^3/s = 0.6463 million gallons per day (mgd)
1 mgd = 1.547 ft^3/s

LENGTH

1 inch (in.) = 2.54 centimeter (cm)
100 cm = 1 m
1 m = 39.37 in.
1 m = 3.281 feet (ft)
1 yard (yd) = 0.9144 meter (m)
1,000 m = 1 kilometer (km)
1 km = 1.609 miles (mi)

POWER

1 horsepower (hp) = 0.746 kilowatts (kW)
1 kW = 1.341 hp

PRESSURE

1 lb/square inch (psi) = 2.307 ft of water
1 ft of water = 0.4335 psi
1 atmosphere (atm) = 14.7 psi
1 atm = 29.92 in. of mercury (Hg)
1 atm = 33.90 ft of water
1 atm = 760 millimeters (mm) of mercury

TEMPERATURE

Degrees Fahrenheit (°F) = ($\frac{9}{5}$)°C + 32
Degrees Celsius (°C) = $\frac{5}{9}$(°F – 32)

VOLUME

2 pints = 1 quart
8 pints = 1 gallon (gal)
4 quarts = 1 gal
1 quart = 32 fluid ounces (oz)
1 gal = 128 fluid oz
1 gal = 3.785 liters (L)
1 L = 1.0567 quarts
1 L = 1,000 milliliters (mL)
3,785 mL = 1 gal
1,000 L = 1 cubic meter (m^3)
1 ft^3 = 7.48 gal
1 million gallons (mil gal) = 3.07 acre-ft
1 acre-ft = 325,829 gal

WEIGHT

1 gram (g) = 1,000 milligrams (mg)
1,000 g = 1 kilogram (kg)
1 pound (lb) = 454 g
1 lb = 7,000 grains (gr)
1 kg = 2.205 lb
2,000 lb = 1 ton
1 mg/L = 1 part per million (ppm)
1 grain per gal (gr/gal) = 17.1 ppm

APPENDIX

B COMMON FORMULAS

COMMON CONVERSIONS AND FORMULAS

Acre-feet to cubic feet:

$$(\text{Number of acre-ft})(43{,}560\ \text{ft}^3/\text{acre-ft}) = \text{Number of ft}^3$$

Acre-feet to gallons:

$$\text{Number of gal} = (\text{Number of acre-feet})(43{,}560\ \text{ft}^3/\text{acre-ft})(7.48\ \text{gal/ft}^3)$$

Cubic feet per second to gallons per minute:

$$\text{Number of gpm} = (\text{Number of ft}^3/\text{s})(60\ \text{s/min})(7.48\ \text{gal/ft}^3)$$

Cubic feet per second to million gallons per day:

$$\frac{(\text{Number of ft}^3)(60\ \text{s})(1{,}440\ \text{min})\ (7.48\ \text{gal})\ \ (1\ \text{mil gal})}{\text{sec}\qquad \text{min}\qquad \text{day}\qquad \text{ft}^3\quad 1{,}000{,}000\ \text{gal}} = \text{Number of mgd}$$

Gallons per minute to cubic feet per second:

$$\text{Number of ft}^3/\text{s} = \frac{(\text{Number of gpm})}{(60\ \text{s/min})(7.48\ \text{gal/ft}^3)}$$

Gallons per minute to liters per second:

$$\text{Flow, L/s} = \frac{(\text{Flow, gpm})(3.785\ \text{L/gal})}{60\ \text{s/min}}$$

Gallons to cubic feet:

$$\frac{(\text{Number of gal})(1\ \text{ft}^3)}{7.48\ \text{gal}} = \text{Number of ft}^3$$

Gallons to liters or liters to gallons:

$$\text{Number of L} = (\text{Number of gallons})(3.785\ \text{L/gal})$$

$$\text{Number of gal} = 1\ \text{gal}/3.785\ \text{L})(\text{Number of liters})$$

Gallons to pounds:

$$\frac{(\text{Number of gal})(8.34\ \text{lb})}{\text{gal}} = \text{Number of lb}$$

Grams to pounds:

$$\text{Number of lb} = (\text{Number of grams})(1\ \text{lb}/454\ \text{g})$$

Liters per second to gallons per minute:

$$\text{Flow, gpm} = \frac{(\text{Flow, L/s})(60\ \text{s/min})}{3.785\ \text{L/gal}}$$

Million gallons per day to cubic feet per second:

$$\frac{(\text{Number of mgd})(1{,}000{,}000\ \text{gal})\ (1\ \text{ft}^3)\qquad (1\ \text{d})\qquad (1\ \text{min})}{(1\ \text{mil gal})\ \ (7.48\ \text{gal})(1{,}440\ \text{min})\ \ (60\ \text{s})} = \text{Number ft}^3\text{/s}$$

Parts per million to percent:

$$\%\ \text{solution} = \frac{(\text{Known ppm})(1\%)}{10{,}000\ \text{ppm}}$$

AREA

Area of a rectangle = (Length)(Width)

Area of a circle (tank) = (0.785)(Diameter)^2 or πr^2

$$\text{Area of a trapezoid} = \frac{(b_1 + b_2)}{2}(\text{Height or Depth})$$

CORROSION

Langelier index = pH – pH_s

$$\text{Mils penetration per year} = \frac{(\text{Weight loss})(534)(1{,}000)}{(\text{Coupon area, in.}^2)(\text{Time, hr})(\text{Metal density, g/cm}^3)}$$

DENSITY

Density = Mass/Volume

DOSAGE

Chlorine dose = Chlorine demand + Chlorine residual

Chemical feed, lb/d = (Flow, mgd)(Dosage, mg/L)(8.34 lb/gal)

or rearrange to solve for dosage:

$$\text{Dosage, mg/L} = \frac{\text{Pounds of chlorine}}{\text{(mgd)(8.34 lb/gal)}}$$

$$\text{Pounds per day} = \frac{\text{(mgd)(Dosage, mg/L)(8.34 lb/gal)}}{\text{(Percent purity)}}$$

Used when the purity of a substance or solution is <100%.

$$\text{(mgd)}(x\text{, mg/L)(8.34 lb/gal)} = \text{(mgd)(Dosage, mg/L)(8.34 lb/gal)}$$

$$\text{Dosage, mg/L} = \frac{\text{(mL/min)(1,440 min/d)(Alum, lb/gal)(Percent purity)}}{\text{(mgd)(8.34 lb/gal)(3,785 mL/gal)}}$$

$$\text{KMnO}_4\text{, mg/L} = 0.2\text{(Iron, mg/L)} + 2.0\text{(Manganese, mg/L)}$$

FLOW RATE

Flow = Volume/Time

Q(Flow) = (Area)(Velocity)

$$\text{Flow, gpm} = \frac{2.83\text{(Diameter)}^2\text{(Distance, in.)}}{\text{(Height, in.)}^{1/2}}$$

$$\text{Flow, ft}^3\text{/s} = \frac{\text{Flow, gpm}}{(7.48\text{ gal/ft}^3)(60\text{ s/min})}$$

Example:

Q, flow in ft^3/s = (Area, ft^2)(Velocity, ft/s)

(continued)

$$\text{Weir overflow rate} = \frac{\text{Flow, gpd}}{\text{Weir length, ft}}$$

Flow in a pipe that changes size:

$$(\text{Area 1, ft}^2)(\text{Velocity 1, ft/s}) = (\text{Area 2, ft}^2)(\text{Velocity 2, ft/s})$$

Gallons per minute (gpm) to million gallons per day (mgd)

$$\text{mgd} = \frac{(\text{Pumping rate, gpm})(1{,}440 \text{ min/d})}{1{,}000{,}000/\text{Million}}$$

GALLONS PER CAPITA PER DAY

$$\text{Gallons/Capita/Day} = \frac{\text{Volume, gpd}}{\text{Population served/day}}$$

HAZEN–WILLIAMS

$$\text{C Factor} = \frac{\text{Flow, gpm}}{(193.75)(\text{Diameter, ft})^{2.63}(\text{Slope})^{0.54}}$$

$$\text{Slope} = \frac{\text{Energy loss, ft}}{\text{Distance, ft}}$$

$$\text{HL/1,000 ft} = \frac{(147.85)(\text{gpm})^{1.852}}{(\text{C})(\text{Diameter})^{2.63}}$$

KILOWATT

kW = (Number of hp)(0.746 kW/hp)

METER ACCURACY

$$\text{Meter accuracy, \%} = \frac{(\text{Meter reading, gal})(100\%)}{\text{Actual volume, gal}}$$

MILLIAMP SIGNAL

$$\frac{\text{mA maximum}}{\text{Maximum possible read}} = \frac{\text{Unknown number of mA}}{\text{Current read}}$$

MIXTURE

$$\text{Percent mixture strength} =$$

$$\frac{\text{Solution 1, gal (Available \%/100\%)} + \text{Solution 2, gal (Available \%/100\%)}}{\text{gal of solution 1} + \text{gal of solution 2}} \times 100\%$$

$$\text{Percent mixture} =$$

$$\frac{\text{Soln. 1, gal(lb/gal of Soln. 1)(Avail. \%/100\%)} + \text{Soln. 2, gal(lb/gal of Soln. 2)(Avail. \%/100\%)}}{\text{Soln. 1, gal(lb/gal of Soln. 1)} + \text{Soln. 2, gal(lb/gal of Soln. 2)}}$$

$$\times 100\%$$

PERCENT

$$\text{Percent} = \frac{\text{(Low number)(100\%)}}{\text{High number}}$$

$$\text{Average} = \frac{\text{Sum of measurements}}{\text{Number of measurements}}$$

$$\text{Ratio} = \frac{A}{B} = \frac{C}{D} \text{ or } A{:}B = C{:}D$$

PERIMETER

Circumference = π(Diameter) = (3.14)(Diameter)

Rectangle = 2(Length) + 2(Width)

PRESSURE

$$\text{psi} = \frac{\text{Depth, ft}}{2.31 \text{ ft/psi}}$$

$$\text{psi} = (\text{Depth, ft})(0.433 \text{ psi/ft})$$

$$\text{Pressure} = \frac{\text{Force, lb}}{\text{Area, ft}^2}$$

PUMPING

Total head, ft = Total static head, ft + Head loss, ft

$$\text{mhp} = \frac{(\text{whp})}{(\text{Motor efficiency})(\text{Pump efficiency})}$$

where mhp = motor horsepower and whp = water horsepower

$$\text{mhp} = \frac{(\text{Flow, gpm})(\text{TH, ft})}{(3{,}960)(\text{Motor efficiency})(\text{Pump efficiency})}$$

where TH = Total head

Total dynamic head = Suction lift + Pumping head + Friction loss + Velocity head

Net positive suction head = Atmospheric pressure – Static suction lift – Friction head loss – Vapor pressure

Brake hp = (hp)(Motor efficiency)

where hp = horsepower

Water hp = (mhp)(Motor efficiency)(Pump efficiency)

Water hp = (bhp)(Pump efficiency)

Motor hp = bhp/Motor efficiency

Brake hp = whp/Pump efficiency

SEDIMENTATION

$$\text{Detention time, hr} = \frac{\text{Volume, gal}}{\text{Flow rate, gph}}$$

SOLUTION

$(\text{Concentration}_1)(\text{Volume}_1) + (\text{Concentration}_2)(\text{Volume}_2) = (\text{Concentration}_3)(\text{Volume}_3)$ or

$$C_1V_1 + C_2V_2 = C_3V_3$$

Solution or dilution triangle:

Conc. 1	Conc. 2 – Conc. desired	=	No. of parts conc.
Concentration desired			
Conc. 2	Conc. 1 – Conc.	=	No. of parts Total no. of parts

$$\frac{(\text{Number of gal})(\text{Number of parts, conc. 1})}{\text{Total number of parts}} = \text{Number of gallons of conc. 1 required}$$

$$\frac{(\text{Number of gal})(\text{Number of parts, conc. 2})}{\text{Total number of parts}} = \text{Number of gallons of conc. 2 required}$$

$$\text{Percent HTH solution} = \frac{(\text{lb HTH})(100\%)}{(\text{Number of gal})(8.34\ \text{lb/gal})}$$

$$\text{Percent of unknown} = \frac{(x\ \text{lb})(100\%)}{x\ \text{lb} + (\text{Number of gallons})(8.34\ \text{lb/gal})}$$

SPECIFIC GRAVITY

Specific gravity (sp gr) = Density of substance/Density of water

SPECIFIC YIELD

$$\text{Specific yield, gpm/ft} = \frac{\text{Well yield, gpm}}{\text{Drawdown, ft}}$$

or

$$\text{Drawdown, ft} = \frac{\text{Well yield, gpm}}{\text{Specific yield, gpm/ft}}$$

or

$$\text{Well yield, gpm} = (\text{Specific yield, gpm/ft})(\text{Drawdown, ft})$$

TEMPERATURE

°C = $\frac{5}{9}$(°F – 32)

°F = $\frac{9}{5}$(°C + 32)

TOTAL FORCE

Force, lb = (Area, in.^{2})(psig)

VOLUME

Volume of a basin = (Length)(Width)(Depth)

Volume of a basin in gallons = (Length)(Width)(Depth)(7.48 gal/ft^3)

Volume of a cone = $\frac{1}{3}\pi r^2$(Height or Depth)

Volume of a circular tank = πr^2(Height) or Volume of a pipe = πr^2(Length)

Volume of a cylindrical tank = (0.785)(Diameter)^2(Height)

Volume of a trough = $\frac{(b_1 + b_2)}{2}$ (Depth of water)(Length)(7.48 gal/ft^3)

WELL DRAWDOWN

Drawdown, ft = Pumping water level, ft – Static water level, ft

Pumping water level, ft = Drawdown, ft + Static water level

Static water level, ft = Pumping water level, ft – Drawdown, ft

APPENDIX

C

CHEMISTRY TABLES

TABLE C-1 International Atomic Weights (Based on Carbon-12)

Element	Symbol	Atomic Number	Atomic Weight*	Element	Symbol	Atomic Number	Atomic Weight*
Actinium	Ac	89	227.0278	Erbium	Er	68	167.26
Aluminum	Al	13	26.98154	Europium	Eu	63	151.96
Americium	Am	95	(243)	Fermium	Fm	100	(257)
Antimony	Sb	51	121.75	Fluorine	F	9	18.998403
Argon	Ar	18	39.948	Francium	Fr	87	(223)
Arsenic	As	33	74.9216	Gadolinium	Gd	64	157.25
Astatine	At	85	(210)	Gallium	Ga	31	69.72
Barium	Ba	56	137.33	Germanium	Ge	32	72.59
Berkelium	Bk	97	(247)	Gold	Au	79	196.9665
Beryllium	Be	4	9.01218	Hafnium	Hf	72	178.49
Bismuth	Bi	83	208.9804	Hassium	Hs	108	(265.0)
Bohrium	Bh	107	(262.0)	Helium	He	2	4.00260
Boron	B	5	10.81	Holmium	Ho	67	164.9304
Bromine	Br	35	79.904	Hydrogen	H	1	1.00794
Cadmium	Cd	48	112.41	Indium	In	49	114.82
Calcium	Ca	20	40.08	Iodine	I	53	126.9045
Californium	Cf	98	(251)	Iridium	Ir	77	192.22
Carbon	C	6	12.011	Iron	Fe	26	55.847
Cerium	Ce	58	140.12	Krypton	Kr	36	83.80
Cesium	Cs	55	132.9054	Lanthanum	La	57	138.9055
Chlorine	Cl	17	35.453	Lawrencium	Lw	103	(260)
Chromium	Cr	24	51.996	Lead	Pb	82	207.2
Cobalt	Co	27	58.9332	Lithium	Li	3	6.941
Copper	Cu	29	63.546	Lutetium	Lu	71	174.967
Curium	Cm	96	(247)	Magnesium	Mg	12	24.305
Dubnium	Db	105	(262.0)	Manganese	Mn	25	54.9380
Dysprosium	Dy	66	162.50	Mendelevium	Mv	101	(258)
Einsteinium	Es	99	(252)	Meitnerium	Mt	109	(266.0)

(continued)

TABLE C-1 International Atomic Weights (Based on Carbon-12) (continued)

Element	Symbol	Atomic Number	Atomic Weight*	Element	Symbol	Atomic Number	Atomic Weight*
Mercury	Hg	80	200.59	Scandium	Sc	21	44.9559
Molybdenum	Mo	42	95.94	Seaborgium	Sg	106	(263.0)
Neodymium	Nd	60	144.24	Selenium	Se	34	78.96
Neon	Ne	10	20.179	Silicon	Si	14	28.0855
Neptunium	Np	93	237.0482	Silver	Ag	47	107.8682
Nickel	Ni	28	58.69	Sodium	Na	11	22.98977
Niobium	Nb	41	92.9064	Strontium	Sr	38	87.62
Nitrogen	N	7	14.0067	Sulfur	S	16	32.06
Nobelium	No	102	(259)	Tantalum	Ta	73	180.9479
Osmium	Os	76	190.2	Technetium	Tc	43	(98)
Oxygen	O	8	15.9994	Tellurium	Te	52	127.60
Palladium	Pd	46	106.42	Terbium	Tb	65	158.9254
Phosphorus	P	15	30.97376	Thallium	Tl	81	204.383
Platinum	Pt	78	195.08	Thorium	Th	90	232.0381
Plutonium	Pu	94	(244)	Thulium	Tm	69	168.9342
Polonium	Po	84	(209)	Tin	Sn	50	118.69
Potassium	K	19	39.0983	Titanium	Ti	22	47.88
Praseodymium	Pr	59	140.9077	Tungsten	W	74	183.85
Promethium	Pm	61	(145)	Ununnilium	Uun	110	(269.0)
Protactinium	Pa	91	231.0359	Unununium	Uuu	111	(272.0)
Radium	Ra	88	226.0254	Unumbium	Uub	112	(277.0)
Radon	Rn	86	(222)	Uranium	U	92	238.0289
Rhenium	Re	75	186.207	Vanadium	V	23	50.9415
Rhodium	Rh	45	102.9055	Ytterbium	Yb	70	173.04
Rubidium	Rb	37	85.4678	Yttrium	Y	39	88.9059
Ruthenium	Ru	44	101.07	Xenon	Xe	54	131.29
Rutherfordium	Rf	104	(261.0)	Zinc	Zn	30	65.38
Samarium	Sm	62	150.36	Zirconium	Zr	40	91

Source: US Government Printing Office.
*The numbers in parenthesis indicate mass number of most stable known isotope.

TABLE C-2 Common Chemicals Used and Encountered in the Water Industry

Chemical	Chemical Formula	Use	Miscellaneous
Alum	$Al_2(SO_4)_3 \cdot 14(H_2O)$	Coagulant	
Ammonia	NH_3	Check for chlorine leaks	Make chloramine
Bicarbonate	HCO_3		Carbonate hardness
Calcium carbonate	$CaCO_3$	Primary hardness chemical	
Calcium hypochlorite	$Ca(OCl)_2$	Disinfectant	
Carbon dioxide	CO_2	Recarbonation	
Caustic soda	NaOH	Adjust pH	
Chlorine	Cl_2	Disinfectant	
Chlorine dioxide	ClO_2	Disinfectant	
Copper sulfate	$CuSO_4$	Algae control	
Dichloramine	$NHCl_2$	Disinfectant	Formed when ammonia is added to water containing Cl
Ferric chloride	$FeCl_3$	Coagulant	
Ferric sulfate	$Fe_2(SO_4)_3$	Coagulant	
Ferrous sulfate	$Fe_2(SO_4)_3 \cdot 7(H_2O)$	Coagulant	
Fluorosilicic acid*	H_2SiF_6	Fluoridation	
Hydrated lime	$Ca(OH)_2$	Increase pH and alkalinity	
Hydrochloric acid	HCl	Laboratory, cleaning	
Hydroxide ion	OH		Naturally found in water
Hypochlorite ion	OCl	Disinfectant	
Magnesium hydroxide	$Mg(OH)_2$		Formed in the lime–soda softening process
Monochloramine	NH_2Cl	Disinfectant	Formed when ammonia is added to water containing Cl
Nitrate	NO_3		Fertilizer, sewage, natural deposits that are eroded
Quicklime	CaO	Water stabilization, increase pH and alkalinity	Water softening
Ozone	O_3	Disinfectant	
Potassium permanganate	$KMnO_4$	Control taste-and-odor substances	Oxidize Fe and Mn
Sulfuric acid	H_2SO_4	Decrease pH and alkalinity	Water stabilization
Soda ash	Na_2CO_3	Adjust pH	
Sodium aluminate	$Na_2Al_2O_4$	Coagulant	
Sodium bicarbonate	$NaHCO_3$	Increase pH and alkalinity	Water stabilization
Sodium fluoride	NaF	Fluoridation	
Sodium hexametaphosphate	$(NaPO_3)_n \cdot Na_2O$†	Sequestering agent	
Sodium hydroxide	NaOH	Adjust pH	
Sodium silicate	$Na_2O \cdot (SiO_2)_x$	Coagulant	
Sodium fluorosilicate‡	Na_2SiF_6	Fluoridation	
Trichloramine	NCl_3		Formed when ammonia is added to water containing Cl
Zinc orthophosphate	$Zn_3(PO_4)_2$	Forms protective coating	

* Formerly known as hydrofluosilicic acid or "silly acid."
† Typically n = 14.
‡ Formerly known as sodium silicofluoride.

TABLE C-3 Formula Weights

Compound	Weight in Grams	Compound	Weight in Grams
$AgCl$	143.32	$KHC_8H_4O_4$ (phthalate)	204.224
Ag_2CrO_4	331.73	$KH(IO_3)_2$	389.912
$Al_2(SO_4)_3 \cdot 14(H_2O)$	594.35	K_2HPO_4	174.176
$BaSO_4$	233.39	KH_2PO_4	136.086
$CaCO_3$	100.089	$KHSO_4$	136.16
CaC_2O_4	128.100	KI	166.003
CaF_2	78.077	KIO_3	214.001
CaO	56.079	KIO_4	230.000
$Ca(OCl)_2$	142.985	$KMnO_4$	158.034
$Ca(OH)_2$	74.095	KNO_3	101.101
CO_2	44.010	$Mg(OH)_2$	58.320
Cl_2	70.906	$MgSO_4$	120.36
ClO_2	67.452	MnO_2	86.937
Cr_2O_3	151.990	Mn_2O_3	157.874
CuO	79.545	Mn_3O_4	228.812
$CuSO_4$	159.60	$NaBr$	102.894
$NHCl_2$	85.921	$NaCl$	58.443
$FeCl_3$	162.206	$NaCN$	49.007
FeO	71.846	Na_2CO_3	105.989
Fe_2O_3	159.692	$Na_2Al_2O_4$	163.940
Fe_3O_4	231.539	$NaHCO_3$	84.007
$Fe_2(SO_4)_3$	399.87	NaF	41.988
$Fe_2(SO_4)_3 \cdot 7(H_2O)$	525.97	$NaOH$	39.997
HBr	80.912	$Na_2O \cdot (SiO_2)_x$	Variable
$HC_2H_3O_2$ (acetic acid)	60.052	Na_2SiF_6	188.056
HCO_3	61.017	NCl_3	120.366
HCl	36.461	NH_3	17.030
$HClO_4$	100.458	NH_2Cl	51.476
HNO_3	63.013	NH_4Cl	53.491
H_2O	18.015	NH_4NO_3	80.043
H_2O_2	34.015	$(NH_4)_2SO_4$	132.13
H_3PO_4	97.995	NO_3	62.005
H_2S	34.08	O_3 (ozone)	47.998
H_2SO_3	82.07	OH	17.007
H_2SO_4	98.07	OCl	51.452
H_2SiF_6	144.092	$PbCrO_4$	323.2
HgO	216.59	$Pb(NO_3)_2$	331.2
Hg_2Cl_2	472.09	PbO	223.2
$HgCl_2$	271.50	PbO_2	239.2
KBr	119.002	$PbSO_4$	303.3
$KBrO_3$	167.000	P_2O_5	141.944
KCl	74.551	Sb_2S_3	339.68
$KClO_3$	122.550	SiO_2	60.084
KCN	65.116	$SnCl_2$	189.596
K_2CrO_4	194.190	SnO_2	150.689
$K_2Cr_2O_7$	294.184	SO_2	64.06
$K_3Fe(CN)_6$	329.248	SO_3	80.06
$K_4Fe(CN)_6$	368.346	$Zn_3(PO_4)_2$	304.703

Based on US Government Printing Office table of atomic weights (see Table C-1).

APPENDIX D

MAXIMUM CONTAMINANT LEVELS

TABLE D-1 Maximum Contaminant Levels for Inorganic Chemicals

Chemical	MCL, mg/L	MCL, mg/L*	MCL, mg/L*
Aluminum	1.		
Antimony	0.006		
Arsenic	0.05		
Asbestos	7 million fibers per liter		
Barium	1.		
Beryllium	0.004		
Cadmium	0.005		
Chromium	0.05		
Cyanide	0.2		
Fluoride	2.		
Mercury	0.002		
Nickel	0.1		
Nitrate (as NO_3)	45.		
Nitrate + Nitrite (sum as nitrogen)	10.		
Nitrite (as nitrogen)	1.		
Selenium	0.05		
Thallium	0.002		

* Extra columns are for any future changes in the MCLs.

TABLE D-2 Maximum Contaminant Levels for Volatile Organic Chemicals (VOCs)

Volatile Organic Chemical	MCL, mg/L	MCL, mg/L*	MCL, mg/L*
Benzene	0.001		
Carbon tetrachloride	0.0005		
1,2-Dichlorobenzene	0.6		
1,4-Dichlorobenzene	0.005		
1,1-Dichloroethane	0.005		
1,2-Dichloroethane	0.0005		
1,1-Dichloroethylene	0.006		
cis-1,2-Dichloroethylene	0.006		
trans-1,2-Dichloroethylene	0.01		
Dichloromethane	0.005		
1,2-Dichloropropane	0.005		
1,3-Dichloropropene	0.0005		
Ethylbenzene	0.7		
Monochlorobenzene	0.07		
Styrene	0.1		
1,1,2,2-Tetrachloroethane	0.001		
Tetrachloroethylene	0.005		
Toluene	0.15		
1,2,4-Trichlorobenzene	0.07		
1,1,1-Trichloroethane	0.200		
1,1,2-Trichloroethane	0.005		
Trichloroethylene	0.005		
Trichlorofluoromethane	0.15		
1,1,2-Trichloro-1,2,2-Trifluoroethane	1.2		
Vinyl chloride	0.0005		
Xylenes	1.750†		

* Extra columns are for any future changes in the MCLs.

† The MCL is for one xylene isomer or the sum of all the xylene isomers.

TABLE D-3 Maximum Contaminant Levels for Nonvolatile Synthetic Organic Chemicals (SOCs)

Nonvolatile	MCL, mg/L	MCL, mg/L*	MCL, mg/L*
Alachlor	0.002		
Atrazine	0.003		
Bentazon	0.018		
Benzo(a)pyrene	0.000		
Carbofuran	0.018		
Chlordane	0.0001		
2,4-D	0.07		
Dalapon	0.2		
Dibromochloropropane (DBCP)	0.0002		
Di(2-ethylhexyl)adipate	0.4		
Di(2-ethylhexyl)phthalate	0.004		
Dinoseb	0.007		
Diquat	0.02		
Endothall	0.1		
Endrin	0.002		
Ethylene dibromide (EDB)	0.00005		
Glyphosate	0.7		
Heptachlor	0.00001		
Heptachlor epoxide	0.00001		
Hexachlorobenzene	0.001		
Hexachlorocyclopentadiene	0.05		
Lindane	0.0002		
Methoxychlor	0.04		
Molinate	0.02		
Oxamyl	0.2		
Pentachlorophenol	0.001		
Picloram	0.5		
Polychlorinated biphenyls	0.0005		
Simazine	0.004		
Thiobencarb	0.07		
Toxaphene	0.003		
2,3,7,8-TCDD (Dioxin)	3×10^{-8}		
2,4,5-TP (Silvex)	0.05		

* Extra columns are for any future changes in the MCLs.

TABLE D-4 Maximum Contaminant Levels for Radioactive Elements

Constituent	MCL, pCi/L	MCL, pCi/L*	MCL, pCi/L*
Combined radium-226 and radium-228	5		
Gross alpha particle activity (including radium-226, but excluding radon and uranium)	15		
Tritium	20,000		
Strontium-90	8		
Gross beta particle activity	50		
Uranium	20		

* Extra columns are for any future changes in the MCLs.

TABLE D-5 Maximum Contaminant Levels for Microbiological Contaminants

Constituent	MCL	MCL*	MCL*
Total coliform bacteria	Presence of coliform bacteria in less than or equal to 5% of monthly samples		

* Extra columns are for any future changes in the MCL.

TABLE D-6 Maximum Contaminant Levels for General Minerals

General Mineral	MCL	Acceptable Range
Bicarbonate alkalinity	250 mg/L	
Carbonate alkalinity	None	
Hydroxide	None	
Total alkalinity	None	
Calcium	None	
Chloride	None	
Copper	1.0 mg/L	
Foaming agents (MBAS)	500 mg/L	
Iron	0.3 mg/L	
Magnesium	None	
Manganese	0.050 mg/L	
pH	None	
Potassium	None	
Sodium	None	
Specific conductance	900 microhms/cm	
Sulfate	250 mg/L	
Total dissolved solids	500 mg/L	
Total hardness	None	
Zinc	5.0 mg/L	

TABLE D-7 Maximum Contaminant Levels for General Physical Characteristics

General Physicals	MCL	Acceptable Ranges
pH	None	
Turbidity	None	
Total alkalinity	None	
Hardness	None	
Odor threshold	3 TON (threshold odor number	
Color	15 CU (color units)	

TABLE D-8 Secondary Maximum Contaminant Levels

Contaminant	Maximum Level
Chloride	250 mg/L
Color	15 CU (color units)
Copper	1 mg/L
Corrosivity	Noncorrosive
Fluoride	2 mg/L
Foaming agents	0.5 mg/L
Iron	0.3 mg/L
Manganese	0.05 mg/L
Odor	3 TON (threshold odor number)
pH	6.5–8.5 standard pH units
Sulfate	250 mg/L
Total dissolved solids	500 mg/L
Zinc	5 mg/L

APPENDIX E

ABBREVIATIONS

°C	degrees Celsius
°F	degrees Fahrenheit
%	percent
acre-ft	acre feet
avail.	available
bhp	brake horsepower
C	coefficient (Hazen–Williams) or concentration (note context)
cm	centimeter
D	diameter
ft	foot or feet
ft/s	feet per second
ft^2	square feet
ft^3	cubic feet
ft^3/min	cubic feet per minute
ft^3/s	cubic feet per second
g	gram
gal	gallon
GPCPD	gallons per capita per day
gpd	gallons per day
gpg	grains per gallon
gph	gallons per hour

gpm	gallons per minute
gr	grains
gr-eq	gram equivalent weight
hp	horsepower
hr	hour
HTH	high test hypochlorite
hypo	hypochlorite
in.	inch(es)
kW	kilowatt
lb	pound
L	liter
LI	Langelier index
m	meter
M	million, or mole (note context)
mA	milliamp
MCL	maximum contaminant level
mg	milligram
mgd	million gallons per day
mg/L	milligrams per liter
mhp	motor horsepower
mil gal	million gallons
min	minute
mL	milliliter
mol	mole
mpy	mils penetration per year
ntu	nephelometric turbidity units
oz	ounce
pH	hydrogen ion concentration
ppm	parts per million
psi	pressure per square inch

Q	flow
r	radius
req	required
s	second
sed	sedimentation
soln.	solution
sp gr	specific gravity
temp.	temperature
TDS	total dissolved solids
V	volume or velocity (note context)
whp	water horsepower
yr	year
ZOP	zinc orthophosphate

BIBLIOGRAPHY

American Water Works Association, *Guidance Manual for Compliance with the Filtration and Disinfection Requirements for Public Water Systems*, 1990.

California State University, Sacramento School of Engineering, *Water Treatment Plant Operation*, Volume 1, third edition, 1994.

California State University, Sacramento School of Engineering, *Water Treatment Plant Operation*, Volume 2, second edition, 1995.

Funk, Robert A., *Advanced Math Problems for Grade III/IV Water Plant Operators*, Gladstone, Oregon: Operator Certification Training, Inc., second edition, 1996.

Price, Joanne Kirkpatrick, *Applied Math for Water Plant Operators* (Lancaster, Pennsylvania: Technomic Publishing Co.), 1991.

ADDITIONAL RESOURCES

Water Treatment Operator Handbook

This book covers everything water treatment operators need to know to perform their jobs and keep in compliance with changing regulations. Every phase of water treatment operator's job is addressed.

Edition: 2002, Softbound, 252 pp.
ISBN 1-58321-184-5; Catalog Number 20481

Water Distribution Operator Training Handbook, Second Edition

A comprehensive text for water distribution operator training.

Edition: 1999, Softbound, 278 pp.
ISBN 1-58321-014-8; Catalog Number 20428

Water Supply Operators I: Water Sources Textbook

Course textbook teaches the basics of the development of water sources for drinking water, groundwater, surface water, raw water characteristics, and the hydrologic cycle.

Edition: 2003, Hardback, 210 pp.
ISBN 1-58321-229-9; Catalog Number 1955.

Water Supply Operations II: Water Treatment Textbook

WSO Water Treatment provides a valuable introduction to the fundamentals of common water treatment processes and techniques.

Edition: 1995, Hardback, 523 pp.
ISBN 0-89867-789-0; Catalog Number 1956

Water Supply Operations III: Water Transmission and Distribution Textbook

Get comprehensive coverage of the basic principles behind the design, construction, operation, and maintenance of water distribution systems in WSO Water Transmission and Distribution.

Edition: 1996, Hardback, 630 pp.
ISBN 0-89867-821-8; Catalog Number 1957.

Water Supply Operations IV: Water Quality Textbook

Contains information about methods of water quality analysis and drinking water regulations.

Edition: 1995, Hardback, 252 pp.
ISBN 0-89867-804-8; Catalog Number 1958

Water Supply Operations V: Basic Science Concepts and Applications Textbook

Find the practical discussions you need of mathematics, hydraulics, chemistry, and electricity in WSO Basic Science.

Edition: 1995, Hardback, 670 pp.
ISBN 0-89867-796-3; Catalog Number 1959

Basic Chemistry for Water and Wastewater Operators

A basic chemistry primer tailored for operators of drinking water or wastewater systems.

Edition: 2002, Softbound, 178 pp.
ISBN 1-58321-148-9; Catalog Number 20494

Basic Microbiology for Drinking Water Personnel

This book provides clear, short descriptions of the waterborne microorganisms—viruses, bacteria, protozoa, and algae—which either pose a human health threat or contribute to distribution corrosion.

Edition: 2001, Softbound, 85 pp.
ISBN 1-58321-121-7; Catalog Number 20463

Operator Certification Study Guide

Prepares water and distribution operators for certification exams by providing an array of multiple choice example questions on water treatment, distribution operations, and math that operators may encounter on certification exams.

Edition: 2003, Softbound, 252 pp.
ISBN 1-58321-287-6; Catalog Number 20517

For pricing and ordering information, please visit the online bookstore at www.awwa.org/bookstore or call AWWA Customer Service at 1.800.926.7337.

www.ingramcontent.com/pod-product-compliance
Lightning Source LLC
LaVergne TN
LVHW080031210926
848481LV00011B/3159